W. I Chamberlain

Tile Drainage

W. I Chamberlain
Tile Drainage
ISBN/EAN: 9783744690515
Printed in Europe, USA, Canada, Australia, Japan
Cover: Foto ©berggeist007 / pixelio.de

More available books at **www.hansebooks.com**

TILE DRAINAGE,

—OR—

WHY, WHERE, WHEN, AND HOW TO DRAIN LAND WITH TILES.

A PRACTICAL BOOK FOR PRACTICAL FARMERS.

By W. I. CHAMBERLAIN, A. M., LL. D.,
Formerly Secretary of the Ohio State Board of Agriculture, and late
President of the Iowa State Agricultural College.

"Tiles are political economists. They are so many young Americans, announcing a better era and a day of fat things."
—RALPH WALDO EMERSON.

A. I. ROOT, Medina, O.
1891.

CHAPTER I.—INTRODUCTORY.

The Scope of the Book.

This little book on tile drainage is to join the ranks of little books, or practical science-primers, on agricultural subjects, written by such men as Prof. A. J. Cook, A. I. Root, and T. B. Terry, and published by A. I. Root. It will feel honored to appear in such company, and will try very hard to be worthy of it.

Please remember that the book is on *tile* drainage It does not try nor wish to cover all drainage; much less all related subjects. There are several good reasons for this. One is, that the size of the book is fixed by the publisher, and he wishes it to be actually a "primer," both in its size and in the clearness and conciseness of its instruction. Another reason is, that drainage is a *progressive* science and art. Much has been learned in the past hundred years. The older, larger books on drainage contain vast amounts of matter that is as useless now as an eighteenth-century chemistry or a last year's almanac. It is a mercy that the readers of this little book need not plod through it. *Tile* drainage has superseded all other kinds of underdrainage, as, for example, that with poles, rails, slabs, brush, cobble stones, or with the mole-plow. It is immensely better than any of these; more durable, more efficient, and really cheaper in the long run. It is really cheaper, too, than open ditching, except for very large receiving ditches, such, for example, as long, large township or county ditches, to convey the surplus water of hundreds or even thousands of acres of pretty level land, owned by many farmers, in separate farms.

Another reason why this little book does not contain all knowledge on this and related topics is the fact that its author does not possess all knowledge! What he does *not* know is omitted. It would make quite a book—indeed, quite a library.

There are other reasons why he confines this book closely to tile drainage, and even to the best and most recent knowledge and practice on that subject; but these will suffice. Motley, in "The Dutch Republic," says that, when the hated Duke of Alva entered a certain city of the Netherlands, no military salute was fired, as would have been proper; but, instead, a deputation from the city fathers appeared before him saying they had come to give thirty-nine good and sufficient reasons for the failure. First, they had no cannon; second, they had no powder; third—but the duke waived the recital of the other thirty-seven reasons.

Tile drainage, and that, too, with round tiles (round inside and round or octagonal outside), is now admitted by all who are well informed on the subject to be the best and really the cheapest sort of drainage for soils that need artificial drainage. Tiles can be made from the local soil and subsoil at or near all localities that need tile drainage. Properly made, burned, and laid, there is no reason why the tiles and the drains should not last for centuries, while all other kinds of underdrains soon rot, choke, are spoiled by land or water vermin, or for other cause become practically useless. To dwell on any of these kinds of drains, explaining their construction, would be like describing to an inquiring road-traveler the angles, landmarks, and windings of *the wrong road*.

This little book, therefore, will not go into curious "ancient history" on the subject, but from first to last will try to give just so much of present, well-established facts, reasons, and methods, as shall enable the wide-awake, thinking, studying farmer to understand the underlying principles and the best present methods, and enable him to put them into practice.

It may not be out of place for me to say that I have been a practical student of drainage for fully forty years, having dug and laid for my father my first cobble-stone drain forty years ago this spring, and having laid more or less *tile* drains nearly every year for the past 26 years. With my own hand, on my own farm, I have laid nearly fifteen miles of tiles, giving thorough drainage to nearly 65 acres; that is, with main drains wherever they were needed, and with laterals chiefly two rods apart; but on 13 acres, three rods apart; also that I have done this work while in debt, *in order to get out of debt*, with necessarily strict economy, and with real pecuniary advantage. I am not, therefore, likely to advise extravagant or unwise expense. I have also carefully examined the drainage systems of many of the best farms in the land, including those of John Johnston, the pioneer in tiling in America; that of his son-in-law, Robt. J. Swan, both of Geneva, N. Y.; T. B. Terry, of Ohio; Sisson Bros., of Illinois; the Agricultural College farms of Iowa and Michigan, and had charge of the thorough drainage of the new State Fair-grounds, Columbus, O., 90 acres. This little book is not, therefore, a compilation of matter drawn from other books or cyclopædias, but, it is hoped, a clear, concise, and systematic statement of important facts and principles drawn chiefly from its writer's own actual experience or verified knowledge. It is conscientiously written by a practical farmer for practical farmers who really need to "tile-out" certain portions of their farms, and to do it at the lowest cost consistent with accuracy and thoroughness. It is written with the strong hope that it may show more clearly *why we tile at all, where it will pay, and when and how it may be done most economically and best;* also in the hope that it may remove some of the needless difficulties and mysteries thrown around the subject by experts and engineers *in order*, it would seem (alas that it need be said!) "to bring grist to their own mills."

And yet, as will be seen further on, the writer advises the

employment of a civil engineer to establish grades and levels in all complicated cases; also always, if possible, the employment of a real, practical expert at digging, grading, and laying tiles, at least for a few days at the first, until the farmer shall himself learn from him just how to handle the various necessary tools, establish the grade, cut the true groove for the tiles, and lay and cover them properly. That is how I learned.

For quite a number of years I have written considerably upon the subject of tile drainage for various leading agricultural weeklies, chiefly *The Country Gentleman*, *The Ohio Farmer*, *The Rural New-Yorker*, and the *National Stockman and Farmer*. In this little work I shall not consciously borrow, even from my own articles there, except in the way of occasional direct quotations with due credit. Those articles were on particular parts of the subject, and often in answer to specific questions from readers. This primer tries to be a brief but complete and systematic discussion of the entire subject; and it tries to present the underlying scientific principles involved, so clearly that its readers not only may but must understand them.

A single caution: It will be necessary, often, to use the two similar words *tiling* (tile-draining) and *tilling* (cultivating). When the writer uses the first he begs that neither printer nor reader will substitute the other for it, or *vice versa*. It kills the meaning every time. Also the two words *undrained* and *underdrained* look alike, but mean the exact opposite of each other. Do not confound them.

CHAPTER II.

Why do we Tile-drain Land? The Theory.

We drain land to remove *surplus* moisture, and to fit it for tillage and rotation of crops and possible horticulture. If there is no surplus moisture because the land is already underdrained by nature, as in the case of sandy loams with porous subsoil, then it does not need tile drainage. A letter lately received says: "I don't understand. Does *all* land need tile drainage?" This is no worse than Horace Greeley, who evidently believed that "the pen is mightier than the"—plow or spade, and hence farmed and drained chiefly and with best financial results with the former implement! He used to say, in substance, that "whatever land it will pay to till, it will also pay to tile." By no means. You might as well insist on picking off stones where there are none, or grubbing out stumps on prairie land that never saw a tree, as upon removing surplus water down through the soil *by* tile drains from land where nature has already removed it down through the soil *without* tile drains. Indeed, Nature was so liberal with us in the creation of our great national domain that probably present prices of land in Ohio will not justify the *thorough* drainage (*i. e.*, with full system of laterals) of more than a quarter of the farms in Ohio, unless it be for purposes of market-gardening or specially high farming; but a very large portion of the Western Reserve needs it through depressions, "swales," "draws," or "sloughs." It is the *surplus* water that needs to be removed promptly, and it should be removed *down through the soil*, and not along its surface. Each of these points should be clearly understood; and so we will at once consider each separately.

First, then, it is the *surplus* water that needs to be removed to give best results in tillage, crop growth, and rotation. What is *surplus* water? It is the water of heavy rains

8 TILE DRAINAGE.

or snows that soak the soil into mud, filling all its pores full, and standing stagnant on or in the soil. In spring and summer it stands in depressions until it evaporates. In fall and winter it stands and freezes and thaws. In either case it injures or perhaps ruins the crops. Thousands of acres of wheat in Ohio and other States during the past wet winter (1890-91) have been killed thus by too much cold water; and the epitaph might be, "Died of wet feet." "Seeing is believing," and I have seen it from car-windows for hundreds of miles, the dry knolls and slopes having good wheat; the wet depressions or flat surfaces having little or none. Nor does water have to stand *upon* the surface to kill or greatly

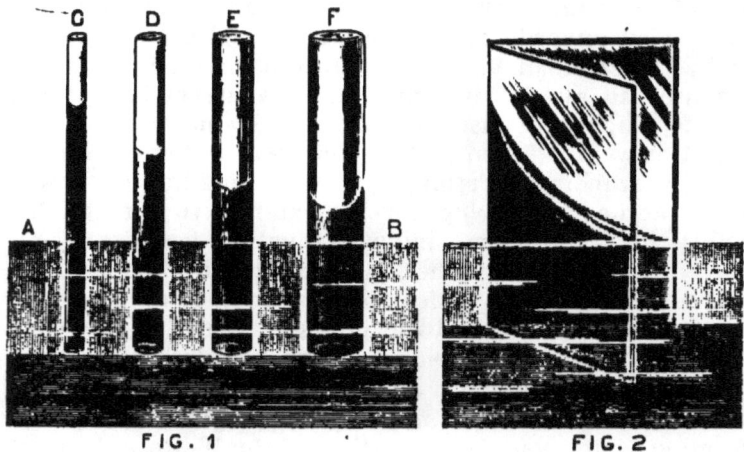

FIG. 1   FIG. 2

Fig. 1.—Capillary attraction in small tubes. The smaller the tube the higher the water rises in it.

Fig. 2.—Capillary attraction between divergent surfaces—of glass for example—opened like the cover of a book.

damage the crops. If it stands stagnant *in* the soil, saturating it, that is, soaking all the pores completely full, it does almost equal damage; for the roots of our agricultural plants need air as well as moisture, and must have it. If long deprived of it they dwindle or even die. The moisture

TILE DRAINAGE.

that they need is held up in the soil by what is called "capillary attraction." Capillary attraction is the force of adhesion between liquids and solids which makes water *rise* in small or "hairlike" tubes; for "capillary" comes from a Latin word which means hairlike. If several small glass tubes, open at both top and bottom, and of different sizes (diameters), have their lower ends standing in a dish of water, then the water will rise in them above the surface of the water in which they stand; and the smaller the tubes are, the higher the water will rise in them. Fig. 1 illustrates this. Let the line A B represent the surface of the water, and C, D, E, and F, represent small glass tubes. The water will rise in them to different heights, proportioned inversely to the diameters, as shown in the figures. The same thing may be shown between plane surfaces close together. Set two panes of glass into a pan of water vertically, with two vertical edges joined and their sides diverging like a thin book set on end and opened just a little. The water will rise between them and be highest close to where the edges meet, as shown in Fig. 2.

The point is, that water *rises* in small tubes or between close surfaces, and *remains above the level of its source*. The tubes need not be round or smooth or straight. They need not be tubes at all, but simply small connected open spaces between some sort of material surfaces. For example, the wick of a lamp is a small bundle of crooked capillary spaces or pores, and the wick lifts the oil in the lamp to the top of the wick. The flame does not "draw" it. It simply consumes it as the capillary attraction lifts it. A fine sponge, with its lower end in water, lifts the water all through its own pores. You pour water into the "saucer" of your flower-pot, and capillary attraction lifts the water (entering at the hole in the pot) all through its soil. In the same way the water in the soil of a field is drawn to or near the surface to supply vegetation in a dry time. It comes from the water stored deeper down in a wet time.

But the soil, like a sponge, has some spaces or pores *too*

large for capillary attraction to hold the water in them up to or near the surface. These larger ones are *filled with air*, and must be, or plants will die. If you plunge a sponge wholly into water (submerge it) these proper air-spaces are filled with water; and if a tenacious clayey soil is not tile-drained, then in a very wet time it is in effect plunged in water. Its needed air-spaces are filled with water. Plant-roots can get no air until *the surplus* water is removed. A soil *full* of water is as unfit for work as a man full of something stronger. A simple experiment proves this. Plant peas or corn in a tight pot or pail, and keep the soil just covered with water. The peas or corn will not grow.

"*Capillary water*" and "*hydrostatic water.*" It will be convenient to use both of these terms frequently in this book, in a technical sense. Let us therefore agree what that technical sense shall be. *Capillary water*, as this little book will use the term, is the water which is brought from below in the soil and subsoil, up to or near the surface, *by capillarity* or capillary attraction; as when a tile-drained or naturally drained soil is *moist* near to the surface long after rain. This is the proper condition for plant-growth.

Hydrostatic water is that which fills the *large* pores of a sponge, for example, when it is submerged in water, and the *large* pores in any soil just after sudden drenching rain, or of an undrained clayey soil, sometimes for days or weeks after heavy rains. Capillary water is a blessing and even a necessity in agriculture. Hydrostatic water, or the water of complete saturation, is a great damage if continued many hours, and a fatal thing to crops if continued many days. It is the purpose and the actual result of tile drainage to remove the hydrostatic water, the damage, and leave the capillary water, the blessing. In untiled clayey soils the hydrostatic water sometimes stands up or nearly up to the surface of the ground for several days after the rain ceases. When every "cradle-hole," or surface depression, even on rolling or sloping land, stands full of water, you may know

TILE DRAINAGE.   11

that the hydrostatic water is at or near the surface all through the soil. When water oozes from the ground, as on slopes, you may know that gravity (hydrostatic pressure) is forcing it out here from a higher level.

In a dry time the hydrostatic water sinks several inches or even feet below the surface in land not tiled or drained by nature. This is because the capillaries have previously carried it to the surface, and the hot dry atmosphere has evaporated and absorbed it, as will be seen presently. You can find just how near to the surface the hydrostatic water stands at any time, simply by digging a hole in the ground. The hole will soon fill up to the top level of complete saturation; *i. e.*, of hydrostatic water in the ground. In a wet time this hole may be a mere post-hole, and may stand full to the very surface. In a very dry time the "hole" will need to be a deep well, perhaps. But the surface-water in each will show the top level of the hydrostatic water that is within reach of hydrostatic pressure; for hydrostatic pressure and capillarity are constantly opposing forces. One pushes the water down; the other lifts it up. One is simply the force of gravity acting on the water. The other is a kind of adhesive attraction between the water and the sides of the pores in the soil. Where these pores are very large, as in gravel or coarse sand, there gravity acts nearly unobstructed, and hence rapidly and over long distances; and so the sand or gravel "veins" or seams in clay subsoils are the "water-bearing strata" for wells and springs, and bring water into them rapidly, and often from long distances. If a tight tube or curb is inserted in the well or spring, the water will quickly rise to the level of its source, even if that is many rods away, and the filtration in such coarse sand is imperfect. In a hamlet in Switzerland, all who drank from a certain spring had typhoid fever. It was traced to the excretions of a typhoid patient, which were thrown out with slops upon the ground, over half a mile distant, and over quite a hill. Proof positive was given by pouring a large quantity of salt brine upon the spot where the excretions had been thrown. The water in the spring soon became perceptibly salt in taste. Fig. 3 illustrates this.

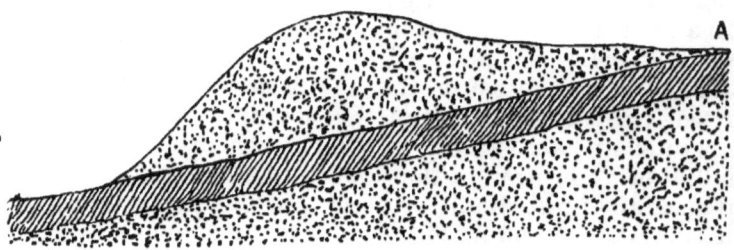

Fig. 3.—Imperfect filtration by coarse sand and gravel as water-bearing strata. The spring that furnished water to the hamlet was at B; the typhoid-soil pollution at A, about half a mile distant. Results as given in the text.

In this way gravel veins in clayey subsoils and artesian wells are explained. The "head" of water is higher, and at a distance, and is connected with the place where the artesian well is dug, by a deep coarse sand or gravel stratum. Fig. 4 illustrates this.

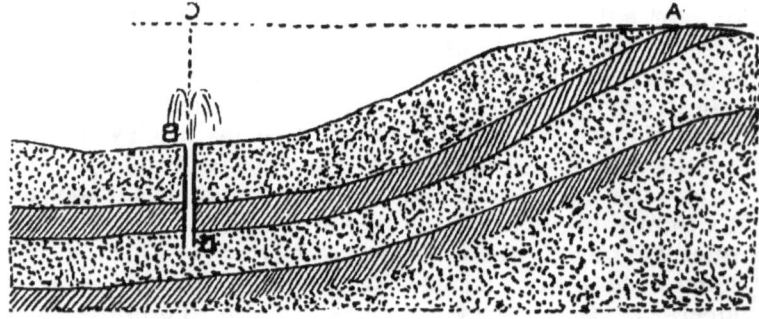

Fig. 4.—Artesian well. A to D is a coarse gravel and sand stratum between nearly impervious strata of clay. Artesian well B D, drilled and tubed down to D. If it is tubed up to C, and the stratum A D is *saturated* clear up to A, then the hydrostatic pressure will force the water in the tube up to a level at C. If tubed only to the surface at B it will make a flowing, or artesian well.

TILE DRAINAGE.

But where the pores in the soil and subsoil are very small, with few of the larger spaces, the hydrostatic pressure is nearly overcome by capillarity, and hence is very slow in its action. This is the case in compact and tenacious clayey soils, especially when not tiled. The water from heavy rains and snows is a long time in soaking into and through the ground, almost making impossible the profitable tillage and growing of root crops and cereals; for, first, the soil can not be tilled early enough in spring nor soon enough after each rain for good results; and, second, the top level of complete saturation (hydrostatic water) is so near the surface of the ground that the plants get little depth of root. On damp clays, maple and apple tree roots run almost or quite on the surface, not for lack of plant food, but from too much water. "Drowned out," we say correctly.

Second. The first point just made, is, that surplus moisture *must be removed.* The second, now to be considered, is, that it must, to secure real success, be removed *down through the soil* and not off along its surface. This is just what underdrainage does for clayey soils and subsoils; and so the full discussion of this point will pretty nearly cover the theory and state the facts of tile drainage; that is, will answer the question that stands at the head of this chapter; viz., "Why do we tile-drain land?"

The point we wish to make is, that the surplus water should be removed *down through* the soil, as by tile drainage, and not *off along its surface,* as where there is no underdrainage natural or artificial. The following are a few of the reasons:

First, because it makes all tillage and harvesting operations easier and more rapid, physically and mechanically. Surface drainage is better than none; but it greatly interferes with all farming operations. If the surface drains are natural, that is, simply made by water action, they will usually be crooked brooks or gulleys, cutting up the field into awkward shapes for cultivation, as in Fig. 5. This represents a ten-acre field of rolling land, which is cut by crooked gulleys or swales into four irregular patches, A, B, C, D, each bounded by the "ragged edge" of a crooked swale, and each requiring nearly as many rounds or turnings with plow, harrow, mower, or twine-binder, as the whole field would take if handled as one unobstructed plat. The sides along the swales, or dry brooks, are wet, crooked, and non-tillable; the angles are far sharper than four right angles of the field, far more vexatious to the farmer, and much more likely, as Dr. Holmes puts it, "to stir up the monosyllables of his unsanctified vocabulary!"

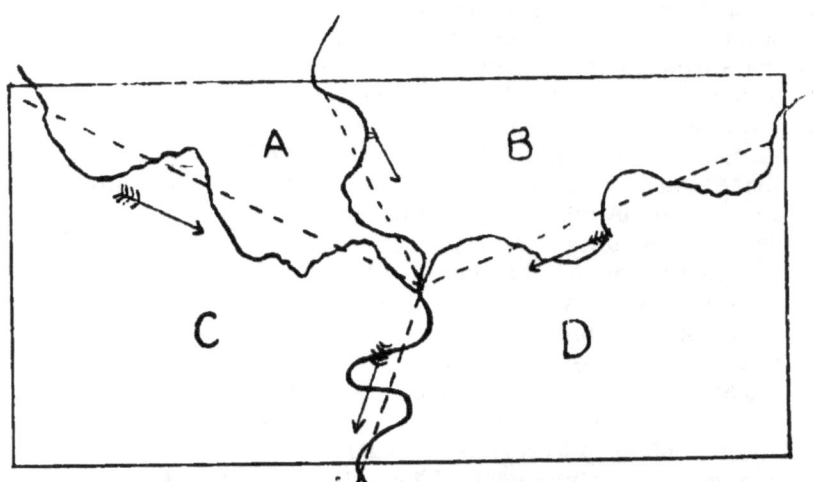

Fig. 5.—Ten-acre field with crooked "swales," or dry brooks, interfering with tillage. For explanation, see text.

Now, if nothing more is done than to cut straight, open ditches, as indicated by the dotted lines, and if only the holes and hollows of the winding brooks are filled and graded (see Fig. 5), even this improves matters considerably. But if *tile* mains are laid where the dotted lines run, and such laterals are put in as the nature of the land requires,

TILE DRAINAGE. 15

then these depressions thus tiled become dry, and fit for tillage in spring time earliest of any part of the field, and soonest after heavy rains, and the whole rectangular field becomes one which it is a delight, and not, as formerly, a weariness and irritation, to farm. I speak from actual experience here as elsewhere.

Second. Removing the surplus water *down through the soil* by means of tile drainage is better than removing it over the top by surface drainage, because the former removes *all* the surplus, not only that on the surface, but that in the soil and subsoil. Even if the surplus is removed from the surface, as it is naturally from rolling clayey lands, while the soil and subsoil are still soaked full, tillage is delayed, and plant-growth is practically suspended, and sometimes the plants actually die. As before remarked, thousands of acres of wheat in Ohio were thus killed during the past winter of 1890.

Third. Removing the surplus *down through the soil* by tile drainage is best because *it prevents loss of fertility* by surface wash. That loss from clayey lands not tiled is sometimes immense. For example, the report of the Ohio Meteorological Bureau shows that nearly *ten inches* of rain and melted snow fell in Hudson during February and March, 1891. But nearly all that amount was *surplus* water, to be removed by some sort of drainage; for the ground was already saturated, too wet, at the beginning of February, and there was very little sunshine or wind to evaporate it, and little growth, even of the wheat, to use the moisture. Indeed, the wheat would have been better without a quarter as much rain. Now, I have noticed the facts on my own wheat, a little over 20 acres. Most of it was top-dressed with about 12 loads per acre, applied on a part of it with the wheat last fall, and on another part plowed under for the preceding crop, and brought to the surface when the land was plowed for wheat. Nearly all the land is thoroughly tiled, with laterals, most of them two and the rest three rods apart. The tiles *handled practically all that water down through the soil*, and left the ground *not at all gullied* by surface wash, and without material loss of fertility from manure or soil. But how fearful would have been the loss and damage except for the tile drains! Think of the immense amount of water that fell on the 20 acres in those 59 days! Almost exactly 50 lbs. on each square foot, or 21,780 *tons* on the 20 acres— enough to take a man and team 3⅛ years to cart away, drawing 21 tons say half a mile each working day! Without the tile drains, all this water must have run off from or remained stagnant on the surface of an already over-saturated soil. If much of it had remained on it, as in case of level land with depressions, it would have wholly ruined the wheat, and terribly "puddled" the soil and damaged its texture. Or if it had run off rapidly, as from quite rolling clayey soil not tiled, it would have gullied the land badly, and caused immense loss of fertility by washing away the soluble elements of plant food and even much of the soil itself. But with the soil thoroughly tiled it did not gully or wash; and the water, filtering down through the soil to reach the drains, nearly three feet deep, left these soluble elements *in the soil* as food for the wheat and future crops. The wheat "survived the flood" in fine condition. The drainage was the Noah's ark that saved both the wheat and the fertility.

Fourth. Removing the surplus water *down through the soil* by tile drainage is best because *it adds* fertility to the soil with each rainfall. That is, it not only prevents loss from surface wash, but it actually secures gain from the rain. Falling rain water, especially during thunder-storms, contains some available nitrogen, small in amount, but valuable. If the land is tiled, the soil acts as a filter, and arrests this fertilizing matter and holds it just where plant-roots can get it.

Fifth. Removing the surplus water *down through the soil* by means of tile drainage helps to *warm* the soil as well as dry it, giving best condition for plant-growth. It warms it

TILE DRAINAGE.

in three ways: It removes the surplus water which otherwise, frozen in winter, must both thaw and evaporate in spring, both of which are chilling processes; second, it keeps the air-spaces open, and warm air ascends in winter from the subsoil, always warmer in winter than the soil. Third, through these open air-spaces warm showers soak down in spring and warm the soil.

As to the first point, thawing and evaporation both keep the soil and the adjacent atmosphere *colder*. Thawing ice, especially by heat from above, is a slow process. Let the water freeze solid in a wooden pump, and try to thaw it with hot water from above, and you will be convinced. In spring the ice on and in an undrained soil must be thawed at an immense waste of sun-heat which would, if the land were drained, be used in warming and drying the soil, and in germinating seeds, in that case already sown. This thawing in spring takes the warmth out of the air just as the ice in an ice-cream freezer, melting by the chemical action of the salt, requires warmth and takes it from the cream, freezing the latter. We all know, too, how chilly the air is when snow and ice are thawing in the fields in spring time. Thus the mere thawing of the ice on and in a saturated clayey soil may delay the starting and growth of crops a full week or more in spring time. But, now, suppose it all to be melted at last. It must still be evaporated from above (nearly all of it in a tenacious undrained clay soil) by the sun's heat before plowing, planting, growth, and tillage can begin; and evaporation is not only a very slow process, but a very cooling one. Wash your hands, even in warm water, outdoors in a brisk wind (where evaporation is rapid), and hold them up in the wind and you will see how quickly they are chilled. We sprinkle pavements and porch floors in summer, and evaporation cools the air. Increased evaporation cools the air after a summer shower. Boys coming out from swimming into a brisk wind are chilled by the evaporation from their wet bodies.

18   TILE DRAINAGE.

Many other illustrations might be given of the fact that both thawing and evaporation are very cooling processes. Undrained clay soils are called "cold soils," and sandy loams are called "warm soils," partly on account of this thawing and this drying by evaporation, that must take place on the former. But tile-draining a clayey soil saves these two wastes of heat, and makes it warm earlier in spring and warmer by several degrees than the undrained adjacent soil all summer, as has been shown by careful tests with thermometers placed in both soils.

The second way in which underdrainage, natural or artificial, warms the soil is by keeping the proper air-spaces (the larger pores) *open* in the soil, and not full of hydrostatic water, or ice in winter. When thus open, the warmth from the warmer subsoil ascends through them and helps to warm the soil. Suppose the soil is frozen. But the subsoil five feet deep will be about $50°$ warm, and will help thaw the frozen soil, which is about $20°$ cooler. If the air-spaces are open this warmth readily ascends.

The third point under this fifth general head was, that tile drainage keeps the pores, or air-spaces, open for the descent of the warm rains of spring and summer, and these carry down their warmth into soil and subsoil. Rain can not descend through the *small* capillary spaces. They are full of moisture held up to or near the surface by capillary attraction. But in the *larger*, proper air-spaces, capillarity does not oppose gravitation, and the warm rain sinks rapidly and warms the soil.

As to frost action, it is a curious and seemingly paradoxical fact that porous soils freeze deeper in winter, though they thaw earlier in spring. Frost goes down faster when the air-spaces are open, just as warmth goes both down and up more readily, as has been seen. If you are skeptical on this point, examine the two kinds of soil described, or try the following simple experiment in winter time. Take two large sponges, as nearly alike as may be. Wet one till its

TILE DRAINAGE. 19

capillaries are full (and its proper air-spaces empty), by simply dipping its lower part in water. *Saturate* the other by submerging it in a small tin pail of water as nearly the size of the sponge as may be. Put the other sponge in the same sort of tin pail, only perforated like a colander. Hang the two pails (containing the sponges) out in the freezing air. You will find the sponge in the colander pail frozen far sooner than the other. It would freeze deeper if long and sunk into the ground. Now set both pails on a gridiron or wire support over a stove where the temperature is not over 60 or 70°, like the spring atmosphere. The aërated sponge in the perforated pail will be thawed long before the other. Air and warmth can circulate through it. It is honey-combed. But warmth gets access to the other sponge only at the bottom, and air only at the top. The ice in an ice-house keeps well if the air is kept out. It melts fast if air gets in and honey-combs it.

Partly as inferences from or corollaries of the proofs given in our fifth proposition above, follow both our sixth and seventh.

Sixth.—Removing surplus moisture down through the soil by tile drainage *lengthens the season* of tillage, crop growth, and harvest. It increases it in spring, as already seen, by saving the time and sunheat otherwise simply wasted in thawing and drying a soaked or flooded soil. It increases it after each soaking rain of the crop season by carrying the surplus water quickly downward (with its warmth) through the open air-spaces, leaving the soil ready for plant-growth, and dry enough for tillage far sooner after each shower. It increases the time of growth and harvest, especially of late varieties of potatoes, by keeping the soil dry enough for growth and digging, even after the heavy rains of autumn begin to come.

Seventh.—Tile drainage *increases the extent of root pasturage.* Roots of most trees (except water-elms, willows, soft maples, and other swamp and lowland trees) and of most agricultu-

20   TILE DRAINAGE.

ral and horticultural crops will not feed below the top level of frequent saturation. Part of my apple orchard is tiled, part not. On the untiled part the roots run close to the surface, get less nourishment, are more disturbed by tillage operations, and over *four times as many* have died out or been killed by tillage as in the tiled part immediately adja-

Fig. 6.—Corn roots, shallow on clayey soil not tiled, in a wet season.

Fig. 7. Corn roots running deep in a tile-drained soil, be the season wet or dry.

cent, while the whole appearance is less thrifty. Two photo-engravings and a diagram in the next chapter illustrate this.

Figures 6 and 7 are fair illustrations of corn growth on tiled and on untiled land. In 6 the line A B is the top level of hydrostatic water—frequent saturation; and in Fig. 7, C D is the line. Now, as the available plant-food of the soil is diffused all through the pores of the soil, and can be taken by the roots only by actual contact, it is plain that an increased depth of root-growth means an increased supply of food for the plant, and the corn shows the results about as shown in Figs. 6 and 7. Tile drainage doubles the depth of the farm; or, as Ralph Waldo Emerson (quoted by French) aptly puts it in an address at Concord, Mass., "This year a very large quantity of land has been discovered and added to the agricultural land, and without a murmur of complaint from any neighbor. By drainage we have gone to the subsoil, and we have a Concord under Concord, a Middlesex under Middlesex, and a basement story of Massachusetts more valuable than all the superstructure. Tiles are political economists. They are so many young Americans, announcing a better era and a day of fat things."

My friend Mr. T. B. Terry and I are both fond of quoting, concerning farming, Daniel Webster's remark about real talent in law practice; to wit, that "there is always room in the upper story." But tile drainage, according to Emerson's apt metaphor, shows to every owner of a clayey farm large areas of valuable unused land; "room in the lower story;" nay, in the very basement!

Eighth.—Removing the surplus water down through the soil by means of tile drainage *helps to disintegrate the soil* and make pulverization possible. I have already spoken of the increased disintegration through deeper frost action; but during the crop season the surplus moisture must be removed before we can "pulverize" with tillage implements. The exact meaning of "pulverize," as its Latin derivation shows, is to reduce to "dust" or fine soft powder. But *surplus* water in the soil *makes mud*, the opposite and enemy of dust. Further, a clayey soil long kept too wet gets "puddled," especially if tramped by live stock or worked with implements when too wet. This means the filling and destruction of the air-spaces, restored only by freezing in ridges, or by tiling and tillage. A puddled clay soil dries into a hard, impervious, brick-like mass. Tile drainage, proper tillage, rotation and care, and keeping live stock off, will prevent this and make a clayey soil more nearly like the fertile sandy and gravelly loams, naturally underdrained and adapted to tillage and rotation of crops.

Ninth.—This brings us naturally to the next proposition; viz., that tile drainage of soils that need it *greatly diminishes the effect of frost in heaving out wheat, clover*, etc , in winter and spring. The action of frost in heaving out roots and plants is powerful and very peculiar, and closely connected with capillary action. How is "hoar-frost" or "stool-ice" formed? Capillary action brings moisture *to the surface* of the ground, but can take it no further, for that is the top of the capillary tubes or pores. It will not run off from the surface, for capillarity is the adhesive attraction between the water and the solid matter (earth) that forms the sides of the capillary pores or spaces, and hence the water can not rise by this force higher than the top of the capillaries that exert the force ; that is, it can not rise above the surface of the ground. Further, the same force that lifts it to the surface holds it there unless some other force takes it away. Two other forces can remove it—frost and warm air ; and the capillaries will, so to speak, pump it to the surface as fast as either of these can take it. Frost takes it up as follows: It first freezes a thin layer of this capillary water at the surface of the ground, and *keeps on* freezing thin layers, each under the bottom of the preceding, each lifting up all previous ones by its own thickness; and so, by morning of a clear cold night, the surface of a damp field will be honey-combed an inch or two deep with this stool-ice or hoar frost.

It is a continuous process or growth from the bottom all night long of loose porous ice. Incidentally it is well to say that, as this hoar-frost lifts with it considerable amounts of surface earth, therefore if clover seed be sown before it is formed, then, when the frost thaws, the seed will usually be nicely covered.

The power of this stool-ice or hoar-frost is very great. If it freezes around and partly under clover and wheat plants, and roots, it will lift them perhaps half an inch each clear frosty night. The sunshine of the next day thaws the frost and leaves the roots lifted just so much; and the next frost will lift them more, and so on until they are out of ground, or so nearly so as to have little vitality for growth. I have seen many fields of wheat on damp undrained soil ruined in this way; and many fields of clover, with the great tap roots sticking up into the air five or six inches all over the fields in spring, like long numb fingers lifted toward heaven, and crying for tile drainage! for tile drainage almost wholly prevents this by *lowering the level of saturation*, or hydrostatic water, so much that only the smallest capillaries can pump it to the surface. Or, aided by the warmth of the sun and the action of the wind, it may even *dry out* a layer of earth at the surface of the ground, so that this may act as a dry mulch, or blanket, a non-conductor or poor conductor of heat and cold between the frost and the top ends of even the smallest capillary pores, and thus nearly or quite thwart the power of the cold atmosphere to form this stool-ice.

These principles and facts throw light upon a curious and hotly contested question in tillage; viz., Shall we retain moisture by cultivating the surface among "hoed crops" immediately after showers in summer? "Yes," say the careful, practical observers, "we know by frequent trials that this really does help to retain moisture in the soil."

"No," say those who have a mere smattering of science; "we stir hay with a tedder to make it dry faster; and if we stir the soil with a cultivator it will just make it dry faster."

So it will *for a time* and *at the top*, until it dries the top half an inch or so and makes it a dry mulch between the hot air and wind and the top ends of the capillaries. For heat and wind, as we have already seen are the other agents besides frost that can take the moisture as fast as the capillary pores can lift it to the surface. But the heat and wind must be able *to come in contact* with the upper ends of the capillaries in order to take their moisture; and this the dry-earth mulch almost wholly prevents; for dry earth is a very poor conductor, both of warmth and air and of capillary water. The fine tillage of the surface destroys the fine small capillary pores *at their top ends*, and substitutes coarser but still very small *air spaces* in the drying earth, which will not readily conduct the water from the capillaries to the top of the ground, nor admit *currents* of warm air to reach the top ends of the capillaries. For, to be more exact, it is not the *heat* or the *wind* that takes up the moisture pumped to the surface by the capillaries, but the air (or atmosphere) itself. This has the physical property or power of absorbing and holding water in large quantities in the form of vapor; and the warmer the air is, the more water (vapor) it will hold. Heat turns the water into vapor, and increases the power of the air to absorb it. Wind aids in absorbing the vapor from the ground, for wind is simply moving air; and when the wind blows, new portions of air pass along and come in contact with the damp surface of the ground and take up or absorb what vapor they can hold; and then they move along and give other drier portions a chance to load up with water —like the line of empty buckets at a fire. filling up at the well or tank or hydrant, and passing on to give place to other empty ones. That is the reason why a windy day is usually a drying one. And so while deep coarse tillage helps to dry out the soil, fine surface tillage after showers does really help retain moisture. The capillaries cease to pump it up, because the wind and heat are kept by this mulch from drinking it up; just as the capillaries of a lamp-wick cease to pump when the flame goes out and ceases to take and use what they pump. Observation attests the fact that fine surface-tillage after showers helps to retain moisture, and a full knowledge of the physical principles involved explains the reason. The sciolists, with their smattering of science, are wrong as usual.

> A *little* knowledge is a dangerous thing;
> Drink deep, or taste not the Pierian spring.

I have tried to explain this point quite fully, partly on account of its bearing on the next two propositions.

Tenth.—Tile drainage on clayey soils *helps the crops to resist drouth better.* It puts these soils in condition to receive this surface tillage sooner after showers, for one thing, and hence to retain moisture better. Again, it permits the crops to be started earlier and pushed faster, and hence makes them more likely to be out of the way of the July and August drouths so common in this latitude. Still further, it puts the whole soil more fully into that loose and spongy condition that enables it actually *to hold more water and yet not be too wet.* A clayey soil not tile-drained tends constantly to settle down and become too compact, especially if worked or tramped when it is too damp; and it almost always *is* too damp in fall, winter, and spring. This reduces the size and number of its capillaries, and hence they can not hold so much; just as a sponge with all its capillaries full loses half its water if compressed to half its size. And this loose, spongy, friable condition of the soil, that fits it to hold the most water in its capillaries, is greatly increased by tile drainage followed by proper tillage. When a clayey soil, not tiled, is thoroughly soaked it becomes more or less puddled and compacted, so that the proper air-spaces become smaller and act as capillaries. It dries only by evaporation from the top, as fast as the capillaries bring the moisture up. But when it finally does get dry it is apt to bake into great dry clods with little moisture, and little chance for roots to penetrate and permeate it; and so a soil that has been rendered light, porous, and spongy, by tiling and proper tillage, really has *more water in it all summer* that is available for plant-growth; that is, more in its *proper capillaries* to start with, and with its air-spaces open. It also wastes less of its moisture by evaporation, for reasons already given, and has its pores open for the quick reception of any sudden showers, thus preventing surface loss from the suddenness and rapidity of the rainfall.

Eleventh.—Tile drainage often, though not always, *diminishes the suddenness and violence of floods.* On this point I have carefully watched the behavior of my own tile-drained land and of adjacent land not drained, and hence have facts as well as theory to offer. I have often noticed this before, but more carefully than ever this last winter of 1890-91. The drained land had its air-spaces open all last winter, and the ground thawed out almost at the beginning of each thaw, and especially of each rain. The rain, descending through these air-spaces, not only thawed the entire ground sooner than if they had been frozen full of hydrostatic water, but at once, often within an hour, set all the tile drains at work. The rain as it fell was received into 30 inches deep of *porous, spongy soil,* and gradually under hydrostatic pressure (acting slowly through the pores) was sent into the tiles. Sometimes the drains would keep right on discharging more and more slowly for several days, and sometimes clear on until the next rain, thus emptying the soil of its surplus, and leaving again this 30 inches of porous soil and subsoil like a great sponge all over the tiles, ready to absorb each rain and make its flow into the open streams below *far slower*, and extended over a much longer time. But the ground adjacent, and not tiled, being soaked full and frozen in December and January, *remained frozen full* nearly all winter, especially where bare or nearly so. All its depressions were puddles of water frozen into solid ice, and the whole surface was thus *without the power of absorbing* or filtering scarcely a gallon of each new rain. Thus the untiled clayey land is like a great flat tin roof from which all the water must run *as fast as it falls*. But the tiled land, especially if covered with a mat of clover or wheat, is like a flat roof covered with a deep layer of fine sponge, which will absorb quite a shower before it will let any run off, and will retain much water, slowly parting with it for hours after the rain ceases.

In early autumn, however, when all the surface depressions are empty on even untiled clayey land, after a long dry summer, we can see that the untiled land would store more water than the tiled before parting with it. But that is the time when disastrous floods do not occur, at any rate. Tile drainage, therefore, after the soil is *saturated* in fall will, as a rule, diminish the suddenness and violence of floods below, just in proportion to the area tile-drained. That is, tile drainage will (and does) furnish a sort of reservoir or storage-layer of porous soil for holding heavy rains back for a time from the water-courses below, thus extending the time and therefore diminishing the violence of the delivery below. Such has been my careful observation of actual facts on and below my own land, and such seems to me to be the rational scientific explanation. And yet I know it is a popular belief that tile drainage increases floods. Open ditches and the clearing of forests do; but tile drainage, I think, does not as a rule.

Twelfth.—Drainage, both open and with tiles, *improves the health of a region*. For best results, large regions must be drained; but even on a single farm it diminishes malarial, diphtheretic, and typhoid tendencies to drain a considerable area around the house and barns by a thorough system of tile drainage. The farmer may not be able to control or greatly influence the general drainage of the township or county in which he lives; but he can control that of the farm on which and the cellar over which he and his family live. In the tenacious and almost impervious clayey soils that are found over so large a portion of Northern Ohio, I believe the cellar should have a good four-inch tile drain laid a foot deep, just inside the wall all around the cellar, covered with coarse gravel, and running under the cellar wall at one corner to a good outlet, and with a good fall. Then a gravel or cement floor will give a sufficiently dry cellar. A wet cellar under the living-rooms is a dangerous thing to health, especially if rubbish or vegetables ever decay there. The cellar should be thus drained first, and then as much of the adjacent yard, garden, and farm as one's means will permit. The pay will come in better health and better crops, and the drainage will in time extend to other farms and to the township and county if they need drainage. Drainage, like charity, "begins at home." The streets of a certain city are said to be marvelously clean because "every man actually sweeps before his own door."

CHAPTER III.

Why do we Tile-drain Land? The Actual Facts; Does Drainage Pay?

The preceding chapter has given theory chiefly, supported by a few facts. According to the theory there given, tile drainage *ought* to bring good results and pay its way on soils that need it, provided the drainage is done economically and the land is properly managed after tiling. What are the actual facts? Has it actually paid where it has been tried? I believe that it has and does, and will in future. As one object of this little book, on the part both of its author and its publisher, is to encourage and promote more successful farming, and as we believe tile drainage is the very basis of successful farming on clayey or low, swampy soils, we shall take considerable pains in this chapter, both by photographic illustrations and by clear statement of facts, to make it plain that *tile drainage actually does pay*, on soils that need it.

TILE DRAINAGE.

I might give facts from many farms in many States—facts of my own observation, and figures given me by the owners. But it may be better to confine myself to my own farm and my own experience, and give that the more fully, both in word and picture.

MY FIRST EXPERIMENT IN THOROUGH TILING.

My first cobble-stone drain I laid for my father about 40 years ago, first in our farm-garden, and afterward in our large garden in Hudson village, where we lived a few years, in order that "we boys" might go through college. Even those single cobble-drains through our gardens, with a good slope, showed such excellent results in mellowing and drying the soil in early spring, and making tillage easier and crops better, that it predisposed me toward underdrainage.

Soon after I bought my father's farm, now nearly 27 years ago, I collected cobble stones for an underdrain through a wet swale or depression in one corner of what is now my orchard. I hired an expert English ditcher, Mr John Winburn, to help me dig and lay the drain. Said he, "Within ten years you will tile-drain this whole field, and will need a four-inch *tile-main* right down this valley, and I would lay it now." I followed his advice. That swale, from being the wettest, became the driest and most friable part of the field, soonest fit to plow in spring, mellowest all summer. A few years after this I lost two crops in succession on that field from excessive wetness—one of potatoes and one of Hungarian grass, both nearly total failures. I at once thoroughly tiled about 13 acres of the 15 (then in orchard) with laterals 33 feet apart and 30 inches deep, and with other four-inch mains as needed, in addition to the one laid before. The next year I harvested 46¼ bushels of wheat per acre from 10 acres of that field, where, bear in mind, I had just had two years of failure. The measurements were certified. Wheat was $1.00 per bushel. The thorough drainage cost me $23 per acre. The first crop of wheat, therefore, *paid twice the total cost of drainage*, besides the increased growth of the apple-trees, and a heavy crop of wheat straw, and as grand a crop of clover and timothy the next year as I ever saw.

Encouraged by these results, and by others later, I went on from year to year increasing the area of thoroughly tiled land, until now I have, as already stated, nearly 65 acres of my total 115, drained with about 15 miles of laterals and the necessary mains.

My very able friend, Mr. B. F. Johnson, of Champaign Co., Illinois, has had much to say for the past ten years in the various agricultural papers for which he has written, of the *evil* effects of tile drainage in Illinois, especially upon orchards, and in increasing the liability to drouth and flood. On these last points, drouth and flood, I have already spoken quite fully in the tenth and eleventh numbers of the preceding chapter. As to general crops and orchards, I have traveled, observed, and inquired quite extensively in nearly all the States of the Union, and have never seen or heard of ill effects of drainage except from Mr. Johnson, and he never offered any ocular or other demonstration. It may be and doubtless is true, as he states, that drouth has increased in Illinois during the past three or four years, and yet the fact be due to other causes than tile drainage. Drouth has also increased during the same period in Central and Western Iowa, where I lived the past five years nearly; but it can not have been caused by tile drainage, for probably *not one per cent of the total area there has been tiled.* Then, too, as to its increasing floods, the Ohio River floods are oftenest quoted. In the first place, the floods *have not increased*, as the statistics of 50 years show; and in the second place, *not one-tenth of one per cent* of the hilly water-shed that causes floods in the Ohio River *has ever heard of tile drainage.* It is simply the old fault in logic, of assigning wrong causes. The logicians call it the fallacy of "*postea, ergo propter*," or "afterward, therefore because of." It rains *after* the barometer

TILE DRAINAGE.

falls and the peacock screams; *therefore* the barometer and the peacock *caused* the rain!

EFFECTS OF TILE DRAINAGE UPON ORCHARD AND WHEAT.

These are best shown from a part of this same field of mine. As already stated, not very long after the orchard was planted, about 13 acres of the total 15 were thoroughly tiled and the other two not. Fig. 8, below, is a diagram

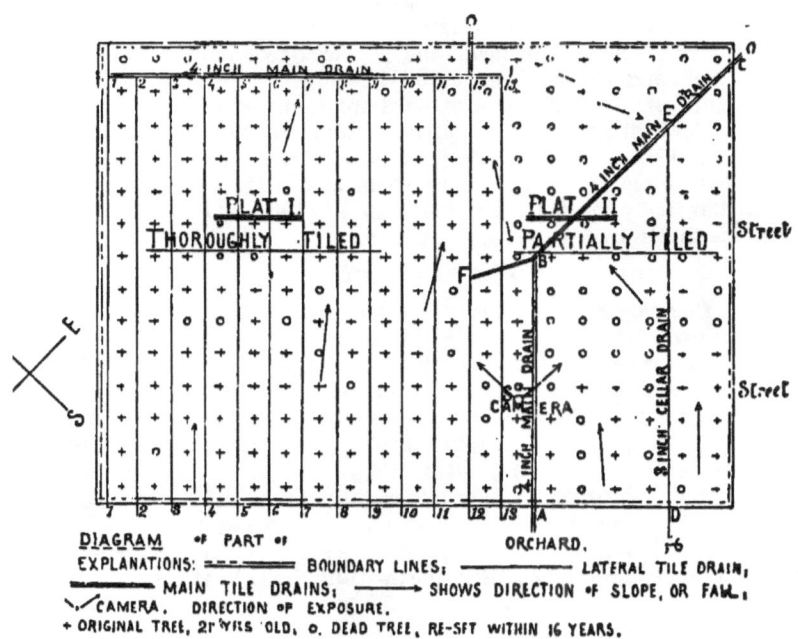

DIAGRAM of PART of ORCHARD.
EXPLANATIONS: ===== BOUNDARY LINES; ———— LATERAL TILE DRAIN; ——— MAIN TILE DRAINS; ——→ SHOWS DIRECTION of SLOPE, OR FALL; ↘ CAMERA. DIRECTION of EXPOSURE.
+ ORIGINAL TREE, 21 YRS OLD, o. DEAD TREE, RE-SET WITHIN 16 YEARS.

showing the exact drainage system of about 6¾ acres of the orchard, including the two acres, or a little more, not thoroughly tiled. The whole plat includes the lowest and most rolling or sloping part of the orchard, and the part not tiled is the most rolling, even of this part, and therefore was left, because I thought it might not need drainage, or at least could be left until I could learn whether so large an outlay paid on the rest of the orchard. At the right of Fig. 8, A B C, with the short spur F B, is the one main drain, and D E is the only lateral in the part not thoroughly tiled. At the left of the figure it is seen that about two-thirds of the plat is thoroughly tiled, having laterals half way between all rows. The rows, and hence the laterals, are 33 feet apart. The part thoroughly tiled is marked "plat 1," and the part only partially tiled is marked "plat 2," and will be so referred to. The mark + shows an original tree, alive and thrifty. The mark O shows one dead, and replaced by a new tree. The photo-engravings, Figs. 9 and 10, show at a glance the results of drainage upon orchard trees and wheat, and show it, as I think, in a most striking and conclusive manner.* Fig. 9 is a photo-engraving of a part of plat 1, and Fig. 10 of a part of plat 2. The camera stood at S, Fig. 9, and pointed diagonally to the left for Fig. 9, and to the right for Fig. 10, as shown by the darts at S, Fig. 8.

THE EFFECTS OF TILING UPON APPLE-TREES.

Compare Figs. 9 and 10. "Seeing is believing," and the camera will not lie. In Fig. 9 the trees are strong and thrifty; their branches meet each other; they shade nearly the whole of the ground; some of the trunks are over 40 inches in circumference a foot above the ground; and the trees are seen, both from the photograph, Fig. 9, and the diagram, Fig. 8, to be nearly all *of the original planting*. But in Fig. 10 the trees are evidently far smaller; there is none of the uniformity, density, and thrift found in Fig. 9; and Fig. 8 shows that more than half of the original trees died; and they have had to be replaced by smaller trees within about 16 years.

---

*With the full consent of the publisher, A. I. Root, I gave in *The Rural New-Yorker* of Aug. 8, 1891, Figs. 8, 9, and 10 (finely reproduced by that paper), and the substance of some six or eight pages which follow here. W. J. C.

TILE DRAINAGE.

ARE THE ENGRAVINGS FAIR AND TRUTHFUL?

Yes, wholly so; or, if any thing, they do not give the facts strongly enough in favor of drainage; for, really, the trees and wheat in Fig. 10 (the same as plat 2 of Fig. 8) have had the best chance. *Naturally* the land in Fig. 10 is better than that in Fig. 9, both for orchard and wheat, being more rolling; and, still further, it lies nearer to the street and to the barn (see Fig. 8), and, to my certain knowledge, has had more manure for the past forty years. Even since I have owned the place, nearly twenty-seven years, it has had most manure, both because one naturally likes to have the crops near the public road look well, and because it is handier to draw the manure there if there is not enough for the whole. Aside from this, which favors the part not tiled, the representation is absolutely fair and truthful.

The point of view was fairly chosen on the line between the drained and undrained part, and in far enough from the southwest edge to give the *middle part* of both plats, unaffected by outside influences, such as prevailing winds, storms, etc. It is true, that some varieties of trees are thriftier growers, such as Baldwins, Astrachans, etc., but the rows of varieties run *across both plats* (Fig. 8); that is, the rows are *perpendicular* to the street, not parallel. The same has been true of the tillage, drilling-in of the wheat, application of superphosphates, etc. All have been perpendicular to the street, and *across both plats*. The same *kinds* of trees (original planting) are in the same relative position in Figs. 9 and 10. Mice, rabbits, and careless tillage, were alike in both plats. In short, every thing is absolutely fair; or, if unfair at all, it is unfair to the tiled part, shown in Fig. 9.

What ailed the trees in Fig. 10? Most of those that died, *died of wet feet.* In spite of the excellent slope, which averages a fall of fully three feet to the hundred, the tenacious and compact clayey soil has been much of the season supersaturated—full of "hydrostatic water." But tree-roots (ex-

Fig. 9.—Growth of apple-trees and wheat in part of orchard tile-drained. For full explanation see the printed matter in the text herewith. Compare trees and wheat here with those in Fig. 10, not tile-drained, all other conditions there being alike or more favorable.

cept of water-growing or swamp-growing trees, willows, white elms, etc.) must have *air* as well as moisture, as was fully explained in Chapter II. When the proper air-spaces in the soil are soaked full of water, the roots suffocate, so to speak, or drown, like land animals held too long under water. And as such animals struggle and swim to the surface to get air, so do the roots run close to or even half upon the surface, *to get air*. Notice the roots, especially of sugar maples and apple-trees, if you have a chance. In my own "sugar-camp," part of the maples grow along a brook-valley on soil rather gravelly and naturally drained. These trees strike their roots down deep, and are very thrifty and strong. But a part of the trees are on wet, cold, level, upland-clay soil. Here the roots run for rods near and even on the surface. I used to think it was because the soil was so thin, and the subsoil so hard and impervious; but upon further study I am convinced that it is because it is too wet. The roots are "swimming on the surface" to get a chance to breathe!

The same is true in this apple-orchard. In plat 2 of Fig. 8 (compare Fig. 10), the roots grow so close to the surface, that, even while the trees were comparatively small, a few years of cultivation cut and broke many of their roots, and the trees died, partly from that cause and partly from overwetness in wet times, and the hard lumpy condition of the soil in dry times.

But in plat 1 (compare Fig. 9) the roots struck down deep into a soil aërated and rendered more mellow by tile drainage, and found sufficient moisture the whole year round, but never excessive, and were helped and not damaged by the tillage of hoed crops growing among them.

EXACT STATISTICS OF MORTALITY.

Fig. 8 gives the means of ascertaining these. As stated, the mark + shows just where each original tree (or tree replaced before the land was tiled) remains alive and thrifty, and the mark O shows each place where one has died and

Fig. 10.—Growth of apple-trees and wheat in part of orchard not tile-drained. Compare with Fig. 9. For full explanation see printed matter herewith in the text.

been replaced since the land was tiled; *i. e.*, within 16 years. It is an exact "graphical diagram," and will repay careful study. Even a glance shows very many more zeros (O) in plat 2 (Fig. 8) than in plat 1. Now count in both, if you please, and you will find that, in plat 2, 49 out of a total 91 have died, or almost exactly 54 per cent. But in plat 1 only 25 out of a total 175 have died, or a little over 14 per cent. That is, *nearly four times* as many in proportion have died in the part only partially tiled as in the part thoroughly tiled. But all other causes, except drainage, have been alike in both, or even more favorable to the part not tiled. We seem to be simply and irresistibly shut down to the conclusion that thorough tile drainage on this clayey soil *has made four times as many trees live*, while a comparison of Figs. 9 and 10 shows that those on the tile-drained soil are far larger and more thrifty. I may add, that, even where original trees are still living, they are not nearly so thrifty or productive, nor is their fruit nearly so nice. For example, I have now, Aug. 10, nearly finished picking and marketing the Red Astrachans on a row which (like all the other varieties) runs *across both plats*. The total yield per tree is fully 50 per cent greater on the tiled part; and in size, beauty, and evenness of shape, there is more than that amount in favor of the tiled.

A still closer examination will show that, even in plat 2, fewer trees died where there was best drainage. For example, in the two rows nearest the street, only 10 out of 28 died, or less than 36 per cent; while in the rest of plat 2, 40 out of 63 died, or over 63 per cent. But virtually these two rows are *one-half* drained; the second from the street, by the cellar-drain, which lies *down the slope*, and hence drains it well, and the first from the street by the deep street-ditch, which is only 30 feet from this row, and averages four feet deep from the level of the trees, and which runs to within three or four trees from the east corner of the plat. The cellar-drain affects the row northwest of itself very little indeed, because it is uphill from the row toward the drain, and a tile drain can not "*draw*" much uphill. The area of very decidedly greatest mortality lies northwest of the cellar-drain, between it and the main drain A B; that is, the part where there are *no drains at all*, except on the outer edges

Still another fact: The row of trees northeast of the main drain H I, in plat 1, has *no lateral drain outside of itself*, northeast; that is, it has only half the benefits of drainage, and that, too, with the main drain H I *uphill* from it. Well, in this row 5 out of 12 trees have died, or over 41 per cent, while in the rest of plat 1 (all thoroughly tiled, with drains on *both sides* of each row), only 21 out of 163 trees died, or less than 13 per cent.

Let me here restate these striking facts in more concise form.

PER CENT OF ORIGINAL TREES THAT HAVE DIED AND BEEN REPLACED WITHIN 16 YEARS.

Where tiled both sides of each row, - - 13 per cent.
Where half tiled, that is, on one side of
 each row, - - - - - 36 to 41 per cent.
Where not tiled at all to speak of, - - 63 per cent.

It is fair to add once more, that the present thrift and bearing capacity of the trees are about in the same ratio.

EFFECTS OF TILE DRAINAGE UPON WHEAT.

As to this point, the companion pictures, Figs. 9 and 10, tell the story. In Fig. 9 (tiled) the wheat is nearly all headed (June 5th of a very late season, 1891); is about four feet high, or up to the waist of a six-foot man, and is thick and thrifty. In Fig. 10 the wheat is a full week later, is just beginning to head; is about two feet high, and so thin you can see the knees and almost the feet of the man who stands in it. The wheat in Fig. 9 at harvest had 33 shocks per acre; and that in Fig. 10, 22 shocks per acre. I gave explicit orders to have the two plats separated in stacking and threshing; but by a misunderstanding of the men it was not done, greatly to my disappointment. The difference, I know,

TILE DRAINAGE.

would have been still greater; this, too, in spite of the fact that the shade in Fig. 9 is far the most dense and most injurious to the wheat yield. In other fields, after more unfavorable winters I have sometimes seen tile drainage more than double the wheat yield, and sometimes even make the entire difference between a failure and a good crop.

Now, the striking facts given above are not of my imagination, nor of my creation, except as I caused them, or, rather, occasioned them, by tiling one plat and not the other. Indeed, I had not noticed the effects so fully before this summer; for, not very long after I tiled the orchard as described, I left the farm for about eleven years, with only an occasional visit. I had even almost forgotten that the land in plat 2 (Fig. 8) was not fully tiled, though I saw in general that the trees and crops were not so good there. But when I returned to my farm this spring, residing on the farm now and managing and working it myself, these facts thrust themselves upon my attention; and the more carefully I examined them the more striking did they appear, and the more surely were they seen to be due to tile drainage I presume that, in all, over fifty gentlemen from various parts of the State and of the United States have visited the farm during the past four months. I have called the attention of all of them to these facts, and all have agreed that they are most striking, and most conclusively in favor of thorough tile drainage, at least for orchards and wheat, on clayey soils, even where quite rolling. I had the photographs taken for Figs. 9 and 10, and others that follow, and now publish them in order that there may be an exact, public, and permanent record of the facts—an ocular demonstration of the effects of drainage.

EFFECTS OF TILE DRAINAGE UPON CLOVER.

Some of my good friends who own and till farms of sandy loam, or limestone "drift" soil, which are naturally drained, fertile, and adapted to clover, wheat, fruits, and root crops,

Fig. 11.—Photograph taken June 5, 1891, of very late season. Clover and timothy were sown together, and exactly alike in Figs. 11 and 12. In Fig. 11 the land is tiled, and clover grows strong and thick and thrifty. Compare Fig. 12, and for further explanation see printed matter herewith.

TILE DRAINAGE.   41

seem to feel and write (in spite of occasional disclaimers) as if clover were a means of *creating* fertility, a sufficient manure in itself to keep up fertility under cropping, and even increase it, without commercial fertilizers or much manure from live stock. I wish they could try my rather cold and naturally unresponsive clayey soil. My chief difficulty has been *to get clover itself to grow well and regularly.* It is a regular and reliable crop with me only on land that has been *tile-drained and considerably fertilized,* or else surface-drained by plowing it in high, narrow "lands," with deep dead-furrows, and *very heavily manured* at some previous time, or directly for the crop with which the clover is sown. No crop in my whole list responds so promptly both to tile drainage and to fertilizers as the clover crop.

Figs. 11 and 12 are companion photo engravings showing the effects of drainage upon clover. Fig. 11 is tile-drained; Fig. 12 is not. In all other respects the treatment was the same, except that the land shown in Fig. 12 lies nearer the street and nearer the barn; has had *more manure in the past,* and has a better slope, easterly, than that in Fig. 11, which is westerly. Both were in winter wheat in the fall of 1889, drilled in the same day or consecutive days; both were seeded to clover and timothy alike in all respects, and the same day, in March, 1890, and both were mown in September of that fall, and weeds, stubble, clover, and timothy were left as a mulch on the ground. About four quarts of clover and six of timothy per acre were sown on both alike. The clover in Fig. 12 came up pretty well, but *did not stand the first winter.* In Fig. 12 there is not enough clover to prevent the hay from being sold as clear timothy. In Fig. 11, especially where superphosphate was used, it was nearly clear clover; that is, the clover was so heavy that it lodged, over much of the field, and virtually smothered the timothy. If clover is a good thing to help improve land, and it unquestionably is, then we owners of clayey farms must tile-drain our land to *get* clover as a reliable crop, in order to improve our land

Fig. 12.—Photograph taken same day with Fig. 11. Timothy and clover sown (March, 1890) exactly as in Fig. 11. Land in Fig. 12 not tile-drained, and clover did not catch so well, and virtually all of it "died of wet feet," or winter-killed during winter and spring of 1890-91. For further explanation see printed matter herewith.

thereby. On the undrained part of the plat, Fig. 12, we got very little clover, even with the use of superphosphate—not enough, as remarked already, to prevent the timothy from being sold as "clear timothy" in any city market. In short, *the tile-drained land had good clover, and that not tiled had almost none*, as seen in Figs. 11 and 12. The photographs were taken June 5th. Neither the timothy nor the clover had begun to head, as they had been kept back greatly by the late and very disastrous May frost.

EFFECTS OF TILE DRAINAGE ON THE VALUE OF MANURES AND FERTILIZERS.

In Chapter II., numbers 3 and 4, I have given the theory, and the reason why tile drainage *should* increase the good effects of manures and fertilizers. I now give a few facts.

In the fall of 1889 I used $5.25 worth of best superphosphate per acre, on about 5 acres of wheat, and none on another strip, side by side, of similar land and treated alike. About a third of each strip was *not then tile-drained;* the rest was. The fertilizer increased the wheat 11 bushels per acre, giving 36 bushels per acre against 25 on the untiled; that is, it paid twice its cost the first crop. But the point now is, that *you could see little benefit on the untiled end of the strip;* but on the tiled end I should judge the yield was doubled; while on the whole it was increased not quite 50 per cent. This year I cut and raked and drew the hay on the two strips separately. Five acres of the unfertilized gave 12 large loads, and the fertilized five acres (10x80 rods) gave 16 large loads, as nearly the same size as the men could make them. But the *gain* on the *untiled* part (one-third the whole, and clear timothy) was only one load, while the gain on the *tiled* part (two-thirds of the whole and nearly clear clover) was *three* loads instead of *two*, as it should have been to be in the same ratio with the other.

I could give other examples, but this will suffice. One great advantage of tile drainage on clayey soils is, that it

Fig. 13.—Timothy on tiled land. Compare with Fig. 14, and see explanation in the printed matter herewith.

Fig. 14.—Timothy on land not tiled. Compare with Fig. 13, and see full explanation in the printed matter herewith.

46   TILE DRAINAGE.

makes manures and fertilizers *tell* so much better. It is most fortunate for us owners of clayey farms, that good commercial fertilizers give such remarkable results on them as soon as they are tiled.

EFFECTS OF TILE DRAINAGE ON THE PERMANENCE OF FERTILIZERS AND CROPS, AND UPON WEEDS.

The companion photo-engravings, Figs. 13 and 14, illustrate this. The photographs were both taken June 5, 1891, before the timothy had begun to show any signs of heading. In Fig. 13 the land is tiled; in Fig. 14, not. All other conditions were alike, except, as in all the other cases, that the untiled part lies nearest the street and the barn, and *has had most manure* in the past 30 years, to my certain knowledge, and has a better slope. Now examine the engravings. In Fig. 13 the timothy is very dense and thrifty, about two feet high, and scarcely a weed or plantain could be found. In Fig. 14 the timothy is thin, scarcely covers the ankles, and is full of weeds, especially of that miserable pest of thin, wet soils, the broad-leaved plantain, large numbers of which are clearly seen in the engraving, or, at least, in the original photograph. The two photographs are of the two parts (end to end) of one strip. Both were sown to timothy and clover mixed, in the growing wheat, six years ago last spring, with superphosphate and fine, pure bone meal, but no manure. In two years most of the clover had "gone out," especially on the part not tiled, Fig. 14. Gradually the timothy has thinned out, too, on the undrained part, and weeds and plantain have come in in its place, and crowded out still more of the timothy. This year the tiled part had *fully twice as much hay per acre*, and it was as fine timothy as you or I ever saw—40 big loads (cocked over night, and compact) on the whole ten acres. On the tiled part the timothy was as heavy as ever before—as heavy as I ever saw, and it seems almost a sin to plow the sod up for wheat, as I am now doing (Aug. 10, 1891), simply because it has been "down" in grass

Fig. 15.— Effects of tile drainage of the land upon the barn-room of the farm. The 100-ton barn must be enlarged for the 75-acre farm. See printed matter herewith.

so long, and because I have enlarged the field from adjacent pasture land, and want it all alike, to start a regular rotation by plats.

EFFECTS OF TILE DRAINAGE ON BARN ROOM.

Seventeen years ago last spring I built a bank barn out of four old ones, 39x72 ft., with some additions. The hay runs to the basement floor in 24x44 ft. of it, and it is quite roomy, holding 100 tons of hay by cubic contents, though the eaves-posts are only 14 feet above the top of the basement in front, and 7 feet in the rear. When I built it, both my neighbors and myself thought it was abundant, and more too, for all the hay and grain I could raise on my little farm of 126 acres, 26 of it so shaded by orchard and maple grove as not to give full crops or pasture. But I have since sold 11 acres of my meadow and rotation land (leaving only 65 now arable, including orchard). Fig. 15 shows how very much too small the barn already is for my crops. The barn is crammed full from basement floor to ridge-pole; gradually filled, settled and refilled during four weeks of haying and harvest, and there are (see Fig. 15) four large stacks outside, two of grain and two of hay—75 large loads of hay and grain outside, and 95 large loads inside, all from 65 acres of tile-drained land; for the plats in the engravings classed as "not drained," and amounting in all to about 15 acres, are now (August, 1891) nearly all tiled, the work being done last winter and spring too late to have much effect on the crops *this year*, and hence fairly classed as "not drained" in the contrasts.

The effect of tile drainage on *my barn room* has been such as is shown in Fig. 15. I must next spring, if possible, enlarge the barn to nearly twice its present capacity, by *lifting the roof* and not much increasing the ground size of the barn. If I "pull down my barns and build greater" I trust it will not be in the spirit of the man in the Scriptures, who was called "a fool" because he proposed to "take his ease (loaf),

Fig. 16.—Specimen tree of Baldwin apples on tile-drained land.

eat, drink, and *be merry,*" like the pig or the ox ; but in the spirit of thankfulness that even we owners of clayey farms have been given, in those farms, almost a mine of wealth, if we only use the brains given us by Him who gave us such clayey land ; if we use our brains, I say, and develop the latent wealth of our farms *by tile drainage, good tillage, manures, fertilizers, and clover.*

And I have pictured and described thus much of the effects of these five handmaidens of success, upon my own farm, in hopes of persuading more of my brother-farmers on such farms to take the first step toward real success, by tiling each year some small portion, at least, of their farms.

I may add that now, Aug. 10, 1891, the winter apples on the tiled part of the orchard are simply grand, hanging nearly as full as they blossomed, for we sprayed the trees with London purple to kill the codling moths and to prevent " wormy apples; " and though we had the rainiest ten days of the season just during the time of spraying, yet the spraying, as is evident now, did very much good. Fig. 16 gives a "specimen tree" of Baldwins. The photograph was taken about Oct. 1, just before the apples were picked, after this paragraph was first written. I give the figure as an incentive to owners of clayey land, first to tile and then to plant at least a small orchard. The apples do not show so clearly as I hoped ; but when Baldwin limbs hang clear down to the ground, you may know that they are very heavily loaded.*

---

* The proof comes to me Nov. 14. The apples have been picked. This tree, shown in the picture, yielded 15 bushels of beautiful Baldwins. The tree is 17 years old. The whole orchard yielded over 1100 bushels, and scarcely one-fourth of the trees bore.—W. I. C.

CHAPTER IV.

Does Tillage Pay Better than Grazing?

Nearly all that was said in the second chapter was based on the assumption that tillage pays best; that the actual plowing of the ground, and the planting and tillage of crops, at least on a part of the farm, pay better than exclusive grazing with no agriculture or horticulture proper; that is, little *tillage* of farm, or planting of garden, orchard, and vineyard. The assumption has not been in favor of *exclusive* vegetable farming, but of "mixed farming;" not in the sense of confused raising of very many kinds of crops and stock, but in the only proper sense of the term; that is, the judicious blending of one or more kinds of animal farming with one or more kinds of cereal and vegetable farming, both being of kinds adapted to the soil of the farm and the tastes and talents of the farmer, and so arranged as not to bring conflict of work.

It may be well to say a few words directly in favor of this assumption, in addition to the facts given in Chapter III , and to review what is known or believed concerning the agricultural development of the human race through its various stages of growth. Such a review, I believe, will establish the truth that, by far the largest population can be supported on any given area under agriculture and horticulture—that is, the plowing and tillage of farm and garden, and that these are the most profitable.

The first stage in the agricultural development of man as man, we suppose to have been what is called the savage state. Men then had language more or less developed, and some power of making and using rude weapons and implements of the chase. They hunted the forests, fished the streams, and picked berries, nuts, acorns, and the like; that is, they lived upon earth's spontaneous products, chiefly flesh and fish. Under this "forest" life (for savage means "pertaining to the forest") it took perhaps several hundreds or even thousands of acres to give food to a single individual. There was no tillage of cultivated crops, no rearing of domestic animals, and the clothing was made of the skins of wild animals killed in the chase.

Next in development came the barbarous, or nomadic stage, when men began to keep domestic animals, living mainly on their meat, and milk and its products, and clothing themselves with woven wool and goat's hair, with a scanty tillage of roots, cereals, and a few fruits. They lived a nomadic life yet, roaming the open plains and valleys in tribes, pasturing their herds and flocks on the spontaneous vegetation, the stronger tribes getting the richer plains and valleys. There was little tillage of crops as yet, for there was neither individual ownership of land, nor permanence of location, even for the tribe, and the weaker tribe was always liable to be driven away from any crops it might have sown and tilled, just as it was about to harvest and use them. Under this system, if system it might be called, there was an advance over the preceding in the possible population to be sustained on a given area, and a few score or perhaps hundreds of acres would support a single person.

Next came the stage of agri*culture*, properly so called, based on permanent location of the tribe, and finally of the small nation, with allotment of portions of land to individuals for more or less permanent use, either by annual rental, long-time rental, for example 49 years as among the Jews, or by actual sale. This gave individual reward proportioned to individual effort, the only spur ever discovered sufficiently powerful to produce intelligent and persistent effort. Under this spur, vegetables, cereals, and fruits were increased in variety, and improved in quality and productiveness; while the same was true in regard to the various kinds of domestic animals. Upon this basis of a permanent agriculture grew up manufactures and commerce; high civilization became possible, and was gradually realized; while, even with the greatly increased ratio of food consumption, seven or eight acres of land sufficed for the support of each inhabitant. That is the ratio in Ohio to-day, where we have 88 persons to the square mile, or one to each 7¼ acres. Carnivorous animals, that once destroyed the meat supply, and were of no real use themselves to man, are themselves destroyed; and the short-horn beef, fed on purely vegetable food, and weighing 1800 lbs. at three years old, is raised on the product of as few acres as the fox or wolf that weighed a few pounds. Meat became far more abundant per acre, and even per capita; and the vegetable, cereal, and fruit supply became almost infinitely better and more abundant, both by the development of better and more prolific varieties, and the invention of machinery for their tillage and harvest.

The highest stage is that of horticulture—garden culture; that is, of intensive agriculture, with small farms tilled like gardens, as in Holland and Belgium, where nearly all the land is thoroughly cultivated in rotation, with the careful saving of all manure, where human food is less of meat and more of fish, fruit, vegetables, and dairy products, and where (Belgium) there are 497 people to the square mile, and about an acre and a quarter feeds, clothes, and shelters each inhabitant.

This very rapid general survey, which might be greatly widened in scope and filled out in detail, seems to show that the position assumed in Chapter II. is a tenable and true one; viz., that tillage of the soil does pay better, and support a larger population than *exclusive* animal industry, grazing in summer and feeding in winter, with little or no tillage. Nature pays for the increased labor put upon the soil. If any doubt still exists, we may dispel it by comparing those parts of Ohio, for example, where we once had dairying as an almost exclusive specialty, with no plowing and no raising of grain, even for flour for the family or feed for the live stock, with those other regions where the thorough tillage of a large part of each farm is practiced, with some live stock and a rotation of crops. The dairy regions of the Western Reserve are good samples of the former, while Stark and Wayne and the Miami Valley counties are good samples of the latter. Formerly in the dairy regions, as already intimated, scarcely an acre was plowed on each farm; all the flour and even most of the potatoes and vegetables were bought, and not more than 15 to 20 cows could be kept on each hundred acres, and even these at the expense of considerable costly mill feed. Under such exclusive dairying, with permanent pastures and meadows, when prices of dairy products declined soon after the war, very many farmers, failing to make a living, sold their clay farms, usually to their more forehanded neighbors (who thus increased their grazing area), and moved west or into the towns. Houses and barns for a time went to decay on farms thus deserted and massed into larger ones; farms depreciated fully one-half in market value; country schools and churches languished, both for lack of funds and of attendance, and a decrease of rural population and prosperity occurred, not unlike that in England some 300 years ago under the almost exclusive grazing and decadence of tillage induced by the famous "fine-wool craze."

The way out of this for us here has begun to come by a gradual return to or beginning of tillage; for the plow is the forerunner of civilization, and the promoter of agricultural wealth. Under surface drainage by plowing in high narrow lands, and *heavily manuring*, good crops of cereals and vegetables may usually be grown on clayey soils, and the heavy drain be stopped for family flour and potatoes and fruit, and mill feed for the dairy. But the possible area of tillage is very small under this system. I believe that this sort of land, aggregating perhaps a third or a quarter of the area of the State of Ohio, can never be made profitable for *extended agriculture*, general farming, and fruit, grain, and vegetable growing, except by gradually tile-draining a few acres each year, and bringing it into productive shape by manures or phosphates, and by wise tillage, thorough cultivation, and free use of clover. And I feel sure that this land thus treated will soon become equal, or nearly so, to the fertile sandy loams of the Miami and Scioto valleys, and of Stark, Wayne, Richland, and similar counties of Ohio, for the growth of wheat, clover, and general farm crops and fruits. But it takes more skill and patience to manage such land. It takes more manure or fertilizers *at the first;* but their effects are *wonderfully lasting*. But I believe that, without tile drainage as a basis, it will be impossible to make such clayey land fit for tillage, clover and rotation of crops to any wide extent— in fact, fit for any thing but a scanty and non paying agriculture, chiefly grazing. I am fully convinced that, on such lands, tile drainage will pay, provided it is made simply *the basis*, and is followed by a wise use of all the manure that can be made and saved from farm live stock of right kinds and good quality, by a free use of clover, and by such use of high-grade superphosphates as may seem wise on actual and careful experiment with them.

As a rule, superphosphates show far more striking effects on clayey soils that need and have tile drainage than on the more sandy loams, or on the limestone soils of Southwestern Ohio, or the black soils with limestone basis found in the northwestern quarter of the State. I think the moderate use of superphosphates wise upon these clayey soils if the manure supply is short; for, even after they are tiled, they need *added fertility at once* over their whole area to insure strong growth of wheat and clover. And if one is trying to bring under cultivation considerable new areas each year, the possible supply of manure from live stock is not sufficient, and clover as yet needs "a starter" to make it grow rank and strong enough to have of itself much fertilizing value. For the past six or eight years I have found high-grade superphosphates exactly the help I have needed to give the land a start; for it is a large undertaking to lay 15 miles of tiles, reclaiming nearly 65 acres, and fitting it for tillage and rotation of crops, and bringing it up to a high and paying grade of fertility. If I had sooner known the real value of superphosphates on my soil as *plant-food* (not mere stimulant, "whisky"), and as a means of starting successful wheat and clover growing and rotation—if I had known this, I say, fifteen years ago as well as I do now, I could certainly have brought up the farm far more rapidly, and I think, too, with much better net financial results. As it was, I at first bought a good deal of manure from town, and, of course, made and saved all I could. But that was greatly insufficient. A shrewd neighbor, Judge S. H. Thompson, once said to me, "Of course, you raise good wheat. There's just about enough stable manure in Hudson village for one clay farm, *and you buy all of that!*"

As to whether tile drainage followed by tillage has paid on my own farm, which is a fair sample of the non-arable clay farms of the Western Reserve, as to this question I have not now a shadow of doubt. Twenty-six years ago I began with sheep-farming. That did not give work for man *and team*, and a team the farmer must have. But horses will soon "eat their own heads off" if not kept at steady and profitable work.

Then I tried dairying, and, as I did not want my wife to be a dairy drudge. I sold the milk at the cheese-factory, and finally to customers on a village milk-route. That paid better—the last quite well. But still there was not steady and remunerative work for *man* and *team* when milking and delivery were over each day. Then I tried (as already stated) plowing in narrow lands with deep dead-furrows for surface-drainage; but this drained off only the water *on* the surface and near the surface, and frequent crop-failures, partial or entire, followed, especially where I plowed more than I could heavily manure; and clover was very uncertain, as it heaved out badly with frost the first winter. This convinced me that our clayey soils not tile-drained are not fitted for extensive plowing and a successful rotation of crops. So I began *thorough* drainage about 15 years ago, having before that time " tiled out" several " swales," or depressions, and seen the excellent results. As already stated, I thoroughly tiled about 13 acres of my young orchard, laying laterals between all rows of trees, 33 feet apart, and 30 inches deep. This field had been quite well manured in the past—much better than any other equal area on the farm. But still the crops would fail, as already stated. But since the drainage, this land, and all other areas drained and fertilized, have been reliable for tillage and crop rotation.

The big wheat crop on the orchard, to wit, 465 bushels from 10 acres, already mentioned, and the fine clover that followed, were my first rewards for tile drainage. They so encouraged me, that, in 1878, I undertook the thorough drainage of 8¼ acres of smooth meadow, and then of 36 acres of rough and exceedingly unproductive pasture land. It was so thin in soil, and so subject to drouth in May, July, and August, that it was of little real value. The whole 36 acres would hardly give summer pasture to seven cows. Mr. Theodore Clark, of southern Portage County (a fruit and wheat region), came by one day during a drouth, and stopped and said :

" I believe that is the most barren-looking field I ever saw. It looks like Sahara."

And it was true then, 16 years ago. The exclusive and close grazing of this whole region left the soil dry and hot in summer, and showers *could not* seem to fall upon it any more than upon Sahara. They would go around us, following the forests, swamps, and water-courses in all directions from us. In particular on my farm it would not rain, oftentimes, when it rained on a large forest less than half a mile north, and another not half a mile southeast of me.

EFFECTS OF TILING, TILLAGE, AND TREE-PLANTING UPON RAINFALL.

Well, I went ahead, and now I have nearly 65 acres tiled, tilled, and in rotation with fine crops; 15 acres of dense orchard, hedges, and roadside maples, a veritable green oasis where once there was what seemed almost a desert, often, in July, August, and September. Wheat, clover, heavy timothy, dense orchard, maples, and hedges, with their cool and solid green, *invite showers, and they come*. Particularly I notice that the orchard, which extends northerly to my maple grove, while the latter merges into the large forest still further north, owned by several of my neighbors—this dense orchard (with the big green crops) seems to draw the showers southward from the north woods, and northward from the south woods, so that now the showers are solid over the whole farm. I think the orchard well nigh pays for itself in the increased rainfall. My neighbors all about me, too, are tile-draining more, and plowing and cultivating more, and substituting a mixed *agriculture* for the exclusive dairy-grazing specialty that had proved so unprofitable and so drouth-producing; and the whole region is becoming better watered, less subject to drouth in summer, and far more fertile and profitable in crops.

To return to my draining: I had drained all but about 15 acres of the 65, when, in May, 1880, I was unexpectedly made Secretary of the Ohio State Board of Agriculture, and left the farm, and suspended the drainage almost entirely for eleven years. In May, 1886, I was as unexpectedly chosen President of the Iowa Agricultural College. November 13, 1890, I resigned that presidency, and within a week I had begun, and within six weeks nearly finished, the remaining 15 acres.

I have given these personal details, and might give many more, to show why I believe in the thorough, systematic tile drainage of such soils as mine, which by nature are unfit for extended tillage. From past experience and experiments

TILE DRAINAGE.

I am convinced that I can take the average slightly rolling shale-clay lands of Northern Ohio (and similar latitudes and conditions), and, by tiling and proper tillage, and use of clover, manures, and superphosphates, within five years bring such land up *profitably* to a capacity of 30 to 35 bushels of wheat per acre, and clover and other crops in proportion. I have many times had from 30 up to even 46½ bushels of wheat per acre, *on all the land* that was both tiled and manured or superphosphated, and the crops the first three years on the average have paid for the tiling and the fertilizers, and sometimes much sooner than that. I feel sure, therefore, *first*, that the tillage of a part of our farms pays better than exclusive grazing and feeding, on all soils fit for cultivation. *Second*, I believe it pays *to fit* for cultivation by tiling portions of each clayey farm not fitted by nature. Tiling is the first step toward agricultural and financial success on thousands of clayey farms in many parts of Ohio and other States, and especially on the "Western Reserve," most of whose soil is like mine. Still further, I believe that the difference in actual average selling price between these clayey farms and those of the naturally drained kind, of the Miami and Scioto Valley counties, if expended judiciously in tiling, fertilizers, and clover seed, will make these farms pay as well as those, and immensely better than now or formerly under exclusive dairying or sheep-farming.

There are other large areas of a different sort all over the land that need either partial drainage or systematic drainage, but at wider intervals than the compact clays require. First, as to partial drainage: All over the rolling prairies of Iowa and bordering prairie States, and even of the sandy loams, are "swales," or "sloughs," and "cat-swamps," or small wet "pockets" that need perhaps one or two good four-inch drains put through them to make them arable and most productive. Without such tiling they often produce little but swamp grass. They have crooked margins or boundaries, and are a serious hindrance in the tillage and harvest operations in large rectangular fields. The agriculture is crooked, and in patches, where it might otherwise be large and straight and rectangular. I have already spoken (page 14) of the loss resulting in all tillage and harvesting operations with a team, and especially with large team implements like the twine-binder. As to the drainage of such fields. Mr. T. B. Terry says (admirably as usual), in his prize farm report (Ohio Agricultural Report, 1882, page 648), "There may be some difference of opinion as to whether underdraining pays where all the land must be drained; but there can be no doubt that it pays to drain waste places, such as cat-swamps, swales, and low clay spots in otherwise fertile fields. It is more work to cultivate around them than to go right through; or, perhaps, in a dry time we may prepare the ground and sow the seed, only to have it destroyed by water. Thus we have all the work to do and no crop, except, perhaps, flag, wild grass, or frogs." Mr. Terry himself thoroughly drained all such places on his farm, and found that it paid a very high rate of interest on the cost of the drainage. I am amazed that all who, like him, have chiefly fertile, sandy, and gravelly loams, with some clayey admixture (like Stark and Wayne Counties and Miami Valley counties), or who have rich, porous farms of rolling prairie soil, with similar depressions—I say I am amazed that they do not do the small amount of tile-draining necessary to make their farming a delight instead of "a weariness to the flesh." On such farms tile drainage is least expensive of all in proportion to area and benefits derived.

Next best it pays, probably, to tile-drain the black-soil lands, once heavily wooded, for example like those in the great limestone region of Western and especially of Northwestern Ohio. These soils are less porous than the prairie soils last described, but far more porous than the stiff, clayey soils first described, and like my own farm. As a rule these lands are more level than either of the other classes, and most of them were originally timbered. The soil and subsoil are often so porous that the drains will "draw" two, three, or even four rods on each side; that is, the drains can be four, six, or even eight rods apart, and yet drain the land quite thoroughly. Even without tile drainage, pretty good wheat may be raised on such land by plowing it in high narrow lands with deep dead-furrows kept well open; but on this plan there is often a good deal of loss of wheat along the dead-furrows.

Sometimes this kind of land is so level that outlets for tile drains can be obtained only by digging long open ditches, paid for by township or county funds raised on the equalized-assessment plan. After such outlet-ditches are dug it would seem to be the height of folly for the individual farmer to fail to get the full benefit of the big ditch he has helped to pay for, get his pay, I say, by systematically "tiling out" at necessary intervals all the land he intends to till. In such regions they speak of "tiling out" the *land;* they really "tile out" the water, and leave the land fitted for the best agriculture.

It thus seems clear to me that tillage pays better than exclusive grazing and feeding, even if tiling must precede it; partial tiling, as simply in the depressions of soils otherwise naturally drained; or thorough tiling by parallel laterals, though at wide intervals, as in the black soils of the prairie regions, or the limestone regions once timbered; or even where the laterals must be not more than two or three rods apart, as on many of the more compact shale-clays of the Western Reserve. As already intimated in substance, the present prices of these latter lands seem to me to make them far cheaper than the lands of the far West, which are often too arid for successful agriculture year after year; cheaper, too, than the high-priced, fertile, naturally drained lands of Ohio—provided only that these clayey lands be tiled economically and well, and be wisely tilled, fertilized, and cropped thereafter. At all events I should be very slow to sell a farm well located in the intellectual, social, and moral atmosphere of the Western Reserve, and move to any region I know of, and buy a farm, relative prices and advantages being all carefully considered.

CHAPTER V.

Where to Drain.

The facts and principles given in Chapters III. and IV. help us to answer this question so far as it relates to *the general localities* and kinds of land that need and will pay for tile drainage. For if, as I believe, henceforth with our existing and constantly increasing population, and prices of land, the tillage of crops of cereals, grasses, and root crops in rotation, *with live stock as one factor*, will pay better and furnish more food per acre, and give employment and subsistence to a larger population than is possible under exclusive grazing and feeding of live stock without the plow as a factor; — if this be true, then on our clayey farms all those areas should be thoroughly tiled which are needed for tillage under such a system of farming; and certainly on our more sandy loams all "cat-swamps," "swales," and "sloughs" should be "tiled out" which are not only themselves unfit for tillage and crop-bearing during average seasons, but which run diagonally or crooked or scattered through fields otherwise rectangular, and the most of whose area (*i. e.*, of the fields) is *naturally* underdrained and fit for tillage; for such spots and streaks of non-arable land in such fields render all tillage and harvesting operations angular, crooked, annoying, and expensive, as illustrated in Chapter I. by Fig. 5. If you could drum together all these wet spots and streaks, like troops at general muster or on dress parade, and put them in one straight solid strip along one side of the field, or out of sight in some back lot, it would not be so bad; but like our sins, or like the "poor

TILE DRAINAGE. 63

relations" of the rich and the "skeletons in the closets" of the high-born and aristocratic, these "cat-swamps," "swales," and the like, thrust themselves upon our notice every "'bout" we plow, cultivate, or reap. Like the ghost of Banquo they "will not down." But you can "down" them by putting their surplus water "down" some 30 inches into well laid tile drains, and can thus make them the most productive parts of your whole farm.

But the question *where* to drain also includes many practical questions as to the exact location and direction of the main drains and the laterals or collecting drains of a system. It is a good rule here to follow nature, but not blindly nor too closely. Adam was put into the garden "to dress it." Man by his intellect has, or may have, a real though limited dominion, not only over the beasts and birds and fishes, but over the earth and its processes of production. And so while we "follow nature" we should lead her—not follow blindly, but with constant improvements.

THE LOCATION OF MAIN DRAINS.

In general they should follow the "dry brooks;" that is, take the general direction and location which the water takes, in a wet time, to get off from the land; for water, taking its own course along the surface, naturally takes what scientists call "the path of least resistance." But man can improve upon this for railways, roads, tile drains, etc. Railways follow the rivers and creeks up the mountain side, or even on more level land; but they constantly straighten the course and lengthen the curves, and improve the grades by cuts and fills and tunnels so as to get a path of *less* resistance under high speed and with heavy loads for the traffic of the road. Just so the main drains in a system should *in general* follow but straighten the dry brooks of a field, lengthening the curves and correcting the grades and making them uniform. Fig. 5, page 14, illustrates this; also Fig. 8, page 31. No one should undertake to tile a field without carefully noticing the natural courses, for one or two seasons, in high water after heavy rains or thaws.

THE DIRECTION OF THE LATERALS.

The same rule holds in regard to laterals. They should in general follow, but straighten and improve, the courses taken by surface-water in time of freshet or flood. If, however, the slope is not very rapid, convenience may lead us to vary the laterals slightly from the direct line down the slope. In my own first thorough drainage, convenience amounting almost to necessity seemed to demand this. The field shown in Fig. 8 was set out with rows of young apple-trees as an orchard. These rows, of course, for convenience, were set parallel to each other and to the sides of the field. But the exact slope was slightly angling with the sides, as shown by the darts in Fig. 8. Now, if the laterals had followed this exact course they would have run into or under apple-trees at many points, and so they were laid parallel to the side of the field to which the slope of the field was most nearly parallel, and were laid just half way between the apple-tree rows. This is almost a necessity in an orchard, and is a great convenience, often, in other fields. As I shall show in the chapter," How to Drain," one can economize very greatly in the expensive hand labor of digging, by plowing the field so as to leave a dead-furrow where each drain is to be, or by running two deep furrows, one in the bottom of the other, with a strong team—four horses if necessary—before digging at all by hand. I have often saved much in this way as well as by filling in with the plow, both of which can best be done if the drains are parallel with some side of the field, as the land can be " plowed out " before digging, and " plowed in " after laying the tiles; and so in my own practice I have tried to follow two general rules in thorough drainage—or, rather, one rule, except as modified by a second; to wit:

First, run the laterals as nearly straight down the slope as may be done consistently with the second rule; to wit:

Second, make the laterals parallel with some side of the field, and thus with the direction of plowing and team work.

ARE THESE RULES CORRECT?

I stated them about as above in a recent series of articles in *The National Stockman*, and gave a diagram covering most of the points brought out in Fig. 8. A contributor to that paper soon after criticised the rules and the methods; and as his criticism states a popular belief which I regard as incorrect, I will give that criticism condensed, and also the substance of my reply, with due credit to the paper named:

He says (condensed), "I do not think Mr. C. is correct in draining as the cut represents" (straight down the slope, according to rule 1). "I want the drains to run either crosswise or diagonally from the way the field is to be plowed, and also diagonally down the slope instead of straight. Water naturally runs down hill, and it will not run sidewise in order to get into the tiles. Now, if you plow the field in the same way the drains run, the water will follow the furrows a long distance *before it will find a drain*, after the ground gets full enough of water for it to run. *The sooner you get it to a drain*, the better. I should run the mains diagonally across the field, and the laterals diagonally the other way." This is the substance of the criticism on this point. The entire criticism is based on the assumption that, when rain comes abundantly enough to cause the tiles to flow, the action of the water is as follows: First, it soaks the ground as full as it will hold, and then the surplus water washes *along the surface*, or along the depressions of the furrows, until it comes directly over a line of tiles, and then soaks straight down into it. This is wholly wrong. The water and drains should never act in this way. If the water flows along the surface thus until it comes directly over the drain, and then soaks down into it, first, it will gully and wash the surface, and then it will wear small vertical channels down into the drain, and carry grit down, and soon obstruct the drain. The water should always find the tiles by soaking *directly down* into the soil wherever it falls, and then, under the force of gravity, or hydrostatic pressure, soaking slowly sidewise under ground through the pores of the soil, and entering the tiles at their joints *at the sides and bottom*, not the top. Such is the teaching of all the authorities on drainage. Such is the way the water and tiles actually behave on my farm, according to my careful observations.

If the theory of my critic were correct it should be carried to its full extent, and the drains run square *across* the slope. Then the water could run straight down the slope until it came directly above each drain, and at once sink straight down into it. This, I repeat, is exactly what we *do not* want. One chief object of tile drainage is to prevent surface-wash, and to prevent the water's flowing along the surface at all; to open the large pores of the soil straight down, nearly to the level of the tiles, and keep them open so that, when rain falls or snow melts it may go down these open pores and work off as it descends *gradually sidewise in the soil* to the tiles; this prevents surface-wash, filters the water, and leaves all plant-food in the soil where the roots can get it.

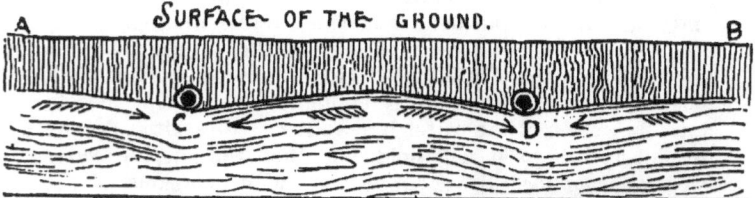

Fig. 18.—Cross-section of tiles properly laid; *i. e.*, straight down the slope, so that the surface *at right angles to the drains* shall be a level line, and the water seek the drains equally from both sides, and the tiles drain the whole ground to uniform depth.

Fig. 18 roughly represents the way water seeks tile drains where they are properly laid — that is, straight down the slope. It gives a cross-section of two tile drains to one who faces down hill. It will be seen that the surface of the ground *crosswise* of each drain will be level; and as the strata of soil and subsoil in bowlder clays that need drainage are usually nearly horizontal (the slopes being made in the past largely by erosion and wash), the water readily follows these strata sidewise, and a little forward down the slope until it finds the tiles and enters them. But, now, suppose the drains

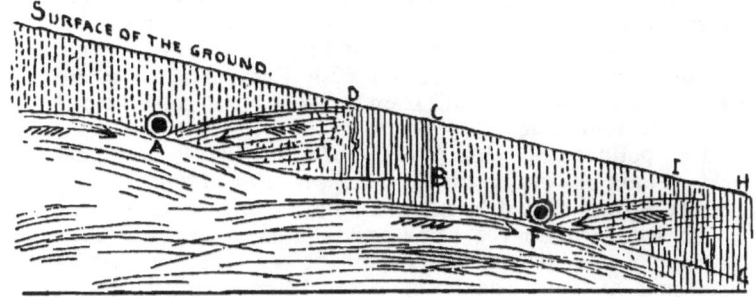

Fig. 19.—Cross-section of tiles improperly laid, crosswise of the slope. Now, as water can soak only *downward* by gravity, and only half way between the drains if laid at the proper distance with reference to the porosity of the soil, it will follow that a considerable part of the land down the slope from each drain (A, B, C, D, and F, G, H, I, in the figure) will not be properly drained. That is, the "suction range" of a tile drain is about as great on level ground as down the slope, while up the slope it has almost none, owing to the opposing force of gravity.

are laid crosswise of the slope, as in Fig. 19. As the soil strata are practically horizontal, if the slope is at all steep then almost no water enters the drain from the downhill side—see Fig. 19. The darts in each figure indicate the lateral flow of the water under hydrostatic pressure from above. But in Fig. 19, only the water between A and D, and that between F and I can enter the two drains *from below*, since water can not flow up hill. This is all wrong. There should be *no down hill and practically useless side to a tile drain;* that is, it should run straight down the slope, and then the water can soak into it equally well from both sides, as in Fig. 18, of course soaking forward down the slope (underground) slightly, rather than at an exact right angle from where it soaks into the surface.

It might be supposed that the water from twice as much space would soak down hill into the tiles (as in Fig. 18), as on level ground; but, practically, no more will do so. The "suction range" of a tile drain depend more upon the porosity of the soil than upon the degree of the slope. Not only is this true theoretically, but I find it practically true, as already stated in connection with Figs. 9 and 10, where we saw that a drain up hill from a row of apple-trees helped the trees very little.

One point more. The plowing of a thoroughly tiled field *should leave no dead furrows* for surface-water to follow. The entire surface should be as level and free from depressions as possible; then the rains and melted snows will soak straight down into the soil, and reach the drains largely by pressing up from the lower level. If you dig a ditch in a wet time you will see the water ooze slowly up from the bottom of your ditch as you dig it—or soon after it is dug—in clear small streams. This shows exactly how and where the water should enter the drains; viz., from below, under the force of hydrostatic pressure. Let me speak still more fully of the porosity of the soil and subsoil as affected by their stratification. Most of our subsoils, and especially the alluvial and drift subsoils, as before remarked, *lie in layers or strata*, nearly or quite horizontal. Examine fresh railway cuts for a few miles through such soils, or dig tile drains for a few weeks in bowlder clay, and you will see. That is, these soils are *more porous horizontally than perpendicularly.* Especially in bowlder clay I have often noticed "seams," layers, or "pockets" of sand or gravel, from which the water bubbles up or seeps in as soon as you remove the imprisoning clay that binds it. Now, although the soils that need drainage are most porous sidewise (or horizontally), yet the perpendicular and diagonal porosity is constantly increased, especially after they are tiled, by the rains soaking down and through to the tiles, and by the roots growing down vertically and diagonally, as well as horizontally, though to less extent horizontally, after drainage than before. And so the most of our subsoils are not, like loose sand or granulated sugar, equally porous in all directions, but are most porous horizontally, and next most porous vertically.

Refer once more to Fig. 18, in which I have tried to illustrate this in a rough way. Suppose tile drains to be located at C and D. Then the rain falling (in a dry time for example) on the slightly uneven surface of the ground all the way from A to B, soaks straight down, at first, at all points on the surface, and not very much sidewise because it falls on all points alike. If the ground is at first quite dry it drinks in the rain rapidly and greedily, not only into the proper air-spaces, but into the larger capillaries, which are now empty for some distance down because the level of hydrostatic water is low from the general dryness If the rain continues long enough, then all the pores, large and small, air-spaces and capillaries, soon become *filled full* up to the level of C and D. Then if the rain continues, as the water can not go down any longer below C and D it will be forced sidewise, horizontally, and sometimes diagonally, by the force of gravity, or hydrostatic pressure, and will enter the tiles and flow off. And the point I wish to emphasize is, that, if the soil acts as it should (the drains being laid straight down the slope), the water will enter the tiles as fast as *all the pores at all points of the surface* can receive and convey it. But if the drains are laid across the slope, and the water is carried down the slope *along the surface* to a line directly over the drain, then it can soak down to the tiles only as fast as the pores in this single narrow strip of soil can receive and convey it down. And so, instead of having the whole area of the field as a filter, you have simply a narrow strip over each tile-drain. The result will be, that, so far as this surface-drainage is successful, *it thwarts the purpose of tile drainage*. The water will, as stated before, even work large holes straight down to the tiles, and you will have little filtering and a final stoppage of the drains. Let me once more emphasize the fact that the water should enter the drains from their sides and bottom, and after filtration through the entire soil, and not straight down from the surface, unfiltered. The latter thwarts the purpose and prevents the benefits of tile drainage.

RAPIDITY OF ABSORPTION AND FILTRATION.

I have spoken of this (Chapter II., section 3), and of the fact that the soil of my farm absorbed and filtered, and my drains carried away, about ten inches of rainfall in February and March, 1891, with no surface wash, gullying, or loss of fertility. I have this week been again reminded of the great absorbing power of a tile-drained soil. We are plowing a 36-acre field for wheat, *across* three plats—one timothy, one wheat-stubble, thick with young clover, and one heavy clover where wheat was harvested (36 bushels per acre) in 1890. Tuesday, Aug. 11, the ground was rather dry to plow —driest in the strong clover turf, and next driest in the timothy turf, dampest in the stubble and young clover. (I had special reasons that justified wheat after wheat, and plowing under the young clover, usually and justly considered bad farming.) Tuesday noon, exactly one inch of rain fell in about an hour, soon followed by 0.26 of an inch; and on Friday by 0.91 of an inch; total, 2.17 inches. The ground took it all in as fast as it came. The first inch did not soak down over three inches into the clover turf not yet plowed, and the whole 2.17 simply soaked down about seven inches and made it scarcely damp enough to plow best. I judge that the soil and subsoil would hold *four inches more*, where tile-drained, before the drains would begin to flow. Soil and subsoil, where drained, are like a vast sponge, quick in absorbing and efficient in retaining the moisture until there is more than the plant-roots can hold, and then transmitting it to the drains, robbed of all its fertilizing matters. The latter are retained *in the soil for the plants.* This 36-acre field is all thoroughly tiled (except between one and two acres recently added from the pasture, and to be tiled this fall)· and at this plowing, no dead-furrows (except three short ones filled nearly full) will be left to encourage surface-wash, and there will be no surface-wash. A plan of the drainage of this field will be given in one of the chapters on " How to Drain."

CHAPTER VI.

When to Drain.

This will be discussed, first, with reference to funds and financial policy; *i. e.*, When can we afford to drain? Second, with reference to economy; *i. e.*, What times of year can we drain best and most cheaply?

First, then, When can we afford to drain? I answer, We *can not* afford *not* to drain if we have land that we need for tillage and rotation, and which is naturally unfit for it, but which can be fitted for it by tile drainage. Shall we wait till we are out of debt and have money to tile it, or shall we tile it in order to get out of debt? The latter, as a rule. If you are buying a clayey farm, buy a smaller one and spend the rest of the purchase money (or debt for purchase) in judiciously tiling all needed for present rotation, and increase the drained area with increasing prosperity. If you already own a clayey farm, sell part of it if you can (as I have lately done 11 acres), and put the money received for it into tiling some of that not sold (as I have just done). I practice what I preach. If necessary, even run in debt cautiously, and as little as possible, in order to tile, and then farm it your very best to pay the debt. This I did while still pretty heavily in debt (mortgage too) for the purchase price of my farm; and I doubt whether I could have got out of debt without thus draining² as a basis for wheat, orchard, clover, and the best tillage of the land and use of the manure from dairy and teams. It helped me to profitable crops of wheat regularly on all land tiled and properly manured or fertilized, and it made the manures and fertilizers go twice as far (into the soil and crops), because it kept them from going half as far (off from the farm in surface wash). This drainage helped me to get profitable crops of wheat and clover regularly, and over constantly increasing areas, instead of occasionally and over small areas heavily manured and surface drained by deep dead-furrows between high, narrow lands as before. It was, I think, the one thing that paid best on my clayey farm, and was the very foundation for the successful tillage and improvement of a soil which is naturally unproductive and even infertile, until it has become one of great productiveness for all crops. Some of these facts in regard to the rewards of tiling on my own land I have already given. And I repeat my belief, that the clayey soils of the Western Reserve, naturally rather infertile except for grass, and naturally irresponsive to tillage, can be made to be really paying farms financially, only by the thorough tile drainage of considerable portions of their area and their careful cultivation in connection with dairying or other keeping of good live stock.

But this does not imply that the work should all be done in one year or even five, or be done by expensive hired labor. A few acres, even one or two, can be tiled each year when other work is light, and when the tiling can be done by the owner himself with the regular farm force.

THE SEASON OF THE YEAR TO TILE.

By November 15th or 20th the usual "fall work" of the farm should be done up. Then if the system of drainage is properly planned and laid out, the work of drainage can go

TILE DRAINAGE.

on all winter, except a few severe days, in latitude 37 to 41°. The winter care of farm stock morning and night will be about all that need interfere with the ditching,

After the fall rains have soaked the earth, but not made it too soft, say about October 1st or 15th, the general location of the mains and laterals should be carefully determined by spirit-level, if the eye and the remembrance of past wet times are not enough ; and the exact location be measured off with tape-line, and indicated by tall stakes, and a deep furrow be plowed with a strong team just where each drain is to be, shading deeper through short knolls, and shallower through each depression, so as to approximate a true grade. A month or so of heavy fall rains will cut and fill in the bottom of these furrows, giving still more nearly a true grade. Then, just before the hand-digging and tile-laying begin, the latter part of November, plow a second furrow in the bottom of each, trying hard to improve the grade as indicated by the cutting and filling of the earth by the rains. My system for nearly three miles of drains, draining about fifteen acres three rods apart, was thus laid out the first week in October, 1890. The second furrows were plowed the third week in November. A day and a half with man and team laid them out and plowed both furrows, making the drains some 8 inches deep. It would have cost some $50 to do this with line and spade, and the *grading* would not have been so good.

The advantage, however, in thus "taking time by the forelock" is not merely the saving in hand-digging, but the fact that we thus have the full preparation for winter ditching made in two days in fall. The loose earth in the bottom of the furrows, and the snow that blows into and lodges in them during even the short flurries of snow, *keep the ground from freezing* too hard to dig all winter long. From Nov. 18 to Dec. 22, last fall there were only two days when we could not dig with comfort, and very few days the rest of the winter and spring up to April first. This utilizes the winter labor of the regular farm force, and makes the actual cash outlay to the average farmer little more than the bare cost of tiles. This answers the question as to when we can dig most economically.

Details of this winter work will be given in the succeeding chapters on "How to Drain," and estimates will be given of the actual cost of the work.

CHAPTER VII.

How to Drain; The Tiles.

Not only have tiles superseded all other kinds of material for drains, but *round* tiles, without collars or joints, have virtually superseded all other shapes of tiles, such as the horseshoe tiles, the sole tiles, the socket or collar tiles, the oval tiles, etc. Cylindrical tiles are cheaper, stronger, better in all respects than any other; and for ordinary drainage there is no need of sockets, joints, or collars. Hence we describe and refer to none but the cylindrical tiles. They may be octagonal outside, without damage, and that form is slightly more convenient in handling, shipping, and laying, because they *lie still better*, and do not roll so much in piling in car or wagon. But this is not important. Buy the fully cylindrical or the octagonal outside and cylindrical inside, as may be cheapest and most convenient. Buy the sort that is made nearest and costs least, if equally good otherwise.

MATERIAL AND HARDNESS OF THE TILES.

Tiles are made of brick clay, and are then called "soft tiles;" also of potters' clay, and are then called "hard tiles." The argument sometimes made for the soft tiles is that they admit the water better. As this is an important matter in drainage, and as I discussed it fully, with account of experiments, in *The Country Gentleman* of April 23, 1891, I will here quote the article entire, with full credit to that paper. I can not state the matter more clearly than I did there.

EXPERIMENTS WITH TILES.

WHERE DOES WATER GET INTO DRAINS?—ABSORPTION BY SOFT TILES DISPROVED;—VALUE OF HARD TILES.

Editors Country Gentleman:—Some six or eight years ago Prof. N. W. Lord and I made some careful experiments with tiles at the Ohio State University, which seemed to prove beyond question that the water enters tile drains at the joints, and not through the pores of the tiles. I think that I gave some account of the experiments at the time in these columns. Since then I have been in the habit of advising the use of hard tiles made of potters' clay, if to be had at the same price, rather than of the soft red tiles made of brick clay, on the ground of greater probable durability, especially near the outlets.

Mr. C. G. Elliott, of Rittman, O., a manufacturer of soft tiles, lately criticised this advice, given by myself, in a western agricultural paper, and said that soft tiles are better because porous, admitting the soil water through their pores into the interior of the drains, as the hard tiles confessedly will not. I therefore gave in reply, briefly, my former experiments and certain new ones made with his own tiles sent me by him as fair samples. They were excellent soft or brick-clay tiles. As I have formerly written to some extent in these columns on drainage, I will now report my recent experiments and conclusions.

First, I took a four-inch tile, medium burned, and set it on end in a deep pail in plaster-of-Paris mortar, and let the plaster harden, inside and outside of the tile. This completely closed the bottom of the tile. I then filled the tile full of water. The water sank perceptibly in the tile with a sort of hissing sound, and the small air-bubbles came to the surface as the pores of the tile greedily drank in the water. In eighty minutes it had sunk two inches, but no water had gone through. I filled it again and left it nine hours. It had then sunk half an inch, *but no water had gone through.* I filled it again and it sank no more. Apparently the same capillary attraction that drank the water into the pores kept it in them, preventing its escape.*

I reversed the experiment with the same tile, still soaked—emptying the water out of the tile and filling the pail all around the empty tile. *No water came through,* and none was absorbed, as the tile was saturated already. I then remembered that water is said to filter through the brick partition of a cistern. The thought suggested itself, that perhaps the water first got through the cracks between the bricks and the mortar, and then with

---

* The reason of this is explained in Chapter II. of this little book.—W. I. C.

water *on both sides* it might possibly go through the brick on that principle of chemical physics known as *osmose* (endosmose and exosmose), by which two liquids on different sides of a bladder or membrane partition mix through the membrane, which is impervious to either alone ! Or, if not on this principle, I thought perhaps the water on both sides of the tile might relieve the retentive force of the capillary attraction and so *let* the water through, while, owing to the greater height on one side, the force of gravity might push it through. The theory seemed all right to explain the alleged facts about cistern filters, but it wouldn't work on the tile. I filled the tile *full* and the pail *half* full, but not a drop would go through, influenced either by gravity, capillary attraction, or the molecular attraction called *osmose*. Not the fraction of a teaspoonful could I get to go through in some twelve hours of time, under all the different circumstances. And I may add that, in the former experiments together with Prof. Lord, we let the water stand for days in a rather damp basement, and not a drop came through. But in a dry wind that evaporated the dampness from the moist surface of the tile, the capillary attraction would supply the place of the moisture thus evaporated.

Now, the tiles experimented with in both these cases were simply medium-burnt brick-clay tiles. But if the land is not drained until the water goes through the walls of the tiles (except in case of flaws or "pin-holes"), it will never be drained ; for a tile, so porous that its walls would suck in one-fifth of its interior contents, was so impervious that it would not let the decimal of a teaspoonful pass through in twelve hours.

I next addressed myself to ascertaining about how fast the water can get in at the *joints*. I set another four-inch tile on end on the one that was closed at the bottom, and was full and saturated ; and I held it firmly down while my man filled it by turning in water rapidly from a full pail ; and when the tile was even full he cried "Now !" and my son and I timed it until it had all run out at the joint—just five seconds — eight gallons per minute ! But Gisborne, one of the earliest and best authorities, figures that, with laterals 36 feet apart, the drains would remove an inch of rainfall in 12 hours if each tile-junction will admit two-thirds of a tablespoonful per minute ! And our tile admitted (or let out) *eight gallons* per minute, under a pressure varying from one foot perpendicular of water down to nothing—several hundred times as fast as need be.

I have written pretty fully, in hopes of exploding the old idea that the water soaks through the tiles. Makers of soft tiles seem still to believe it, and some buyers too. Waring and other authorities on drainage state the case correctly. Waring says ("Draining for Profit and Health," p. 77): "They" — that is, brick-clay tiles—" are porous to the extent of absorbing a certain amount of water, but their porosity has nothing to do with their use for drainage. For this purpose they might as well be of glass. The water enters them, not through their walls, but at their joints, which can not be made so tight that they will not admit the very small amount that will need to enter at each space."

Of the fifteen miles of tile drains on my little farm, nearly half are hard potter-clay tiles, most of them glazed, and about as hard as a jug or earthen crock. They can not and do not crumble or flake with the frost, even at the outlets, where they are constantly freezing and thawing while wet; but the brick-clay tiles at the outlets flake and shell to pieces with the frost. The glazed potter-clay tiles drain the land exactly as well as the porous ones, so far as I can see, for both sorts work perfectly. I have sometimes had the whole bottom course of piles of brick-clay tiles, lying directly on the ground, shell all to pieces in a single winter with our frequent rains and quick freezes and thaws. And so I prefer the hard tiles if they can be had at about the same price. As they are stronger, and burned harder, they can be made thinner and lighter than the brick tiles, and hence cost less for freight and handling. I have just weighed some with the following results:

2-inch, soft, 2 lbs. 13 oz.; hard, 2 lbs. 8 oz.
3-inch, soft, 5 lbs. 13 oz.; hard, 4 lbs. 2 oz.
4-inch, soft, 7 lbs. 6 oz.; hard, 6 lbs. 8 oz.

There is less difference in weight than in size, as the soft ones are lighter in specific gravity; and there is less difference in weight than I supposed until I tested them; also less in proportion in the two-inch ones.

Still, I think the brick-clay tiles will endure for centuries, except at the outlets, if laid below frost, and hard burned. Each one should show clear red color, and give a clear metallic ring when struck with a hammer, or be rejected.

W. I. CHAMBERLAIN.

In brief, my advice is, use hard tiles if you can get them at about the same price as the soft or brick-clay ones. If not, then use the latter, but see that they are well burned, and use the potter-clay, or hard tiles, near all outlets, and for a rod or two back from all outlets. And whichever sort you use, do not delude yourself with the belief that the water enters the drains through the pores of the tiles. It enters at the joints. Sheep will not jump a high fence when a gate stands wide open. Also insist on buying *with the tiles* a sufficient number of "T joints" or "Y joints," or, at least, of tiles with holes cut for joints wherever the water enters the main drain from the laterals.

Fig. 20, page 80, which illustrates many points to be

brought out in the next chapter, shows two T joints and one Y joint near the front end of the stoneboat, as well as round and octagonal tiles, and one socket tile, or sewer-pipe.

CHAPTER VIII.

How to Drain; The Tools; Hand Tools or Machine Tools?

In these days of machinery shall we dig by hand, by horse power, or by steam? After much investigation I am of the clear opinion that the average farmer, on bowlder clay, will dig and fill most cheaply by hand, except the plowing of top furrows, as described in Chapter VI., and the filling by team. While I was Secretary of the Ohio State Board of Agriculture, a field trial of steam and horse power ditching-machines was held on the new State Fairgrounds at Columbus, and later another was held in Marion, Ohio, both under my direction as Secretary. At the two trials, two steam-machines and four or five horse-power ones, tried their powers, and competed for the prizes. At Columbus the ground (common "glacial drift," or "bowlder clay") was so stony that none of the machines did profitable work. At Marion it was far less stony, and one horse-power machine, the "Rennie," of Toronto, Canada, and one steam-power machine, the "Plumb," from Illinois, did excellent and fairly paying work, provided a large job of it were to be done at a time of year when they could be operated. Either of them, and one or two others of which I know, would pay on very large jobs in practically stoneless soil. But none of them will probably pay for the average farmer on bowlder clay, and for the following reasons:

First, the ground is usually too stony. Of the three miles I dug the past winter on my own farm, scarcely a single length of ten rods consecutively could have been dug profitably by any machine I have yet seen, and I have seen nearly all of any reputation. Stones occurred under ground everywhere, from the size of one's fist up to the size of a bushel basket or even a haycock. To the hand-digger they are not a very great hindrance. Those that weigh from fifty to a hundred pounds or so can be removed without much delay. Where they weigh several hundred pounds, or even several tons, one can quite readily find their boundaries and curve the ditch gradually around them. For example, Fig. 20 shows a ditch dug by hand (except top course with plow) 30 inches deep, and with the tiles laid, but not covered. A great bowlder, probably weighing a ton or more, was struck, two feet under ground. When struck with a crowbar the sound revealed its size to an expert ear — too big to be moved, but with the ditch near one side of it. So I just curved the bottom of the ditch out of line about one foot, cutting under the side of the bank and making a total curve some 12 feet long. The water will flow perfectly, and it hindered us so little that we got the usual amount done that day. In fact, we often struck similar ones, both larger and smaller, below the surface, and often had to curve the ditch or spend dollars in removing a great bowlder, only to leave a large deep hole in the bottom of the ditch filled with soft mud, and likely to cause a "sag" in the drain. But such stones, *in such numbers as they usually occur on bowlder clays that need drainage*, make machine digging at any time of the year unprofitable. Every stone makes a long stoppage of four horses and two men, and requires hand work to remove or dig around it.

Second, the machines can not work well on wet ground or in winter, when the top is either muddy or frozen ; but that is just the time when labor is cheapest, when the ground digs easiest, and when the soil water serves to grade the bottom of the ditch.

It will not, at the present stage of invention, pay the average farmer, I think, to own a machine as he does a mower

Fig. 20.—Photo-engraving of tiles, tools, position of workmen, and tiles laid in the bottom of the ditch, and taking a long curve around a very large bowlder in the bottom of the ditch. See printed matter of the text herewith:

or twine-binder. If any one who owns one will dig for you by the rod and board himself, *and furnish hand work to handle the bowlders* that he strikes, and do it cheaper, really, than you can do the work yourself in winter when you would not be earning much otherwise—why, then hire him on a clearly understood contract.

THE HAND TOOLS.

First, there is the ditching-spade, for lifting the *top course* below the furrow. It is shown at the extreme right of Fig. 20, rear end of stoneboat, and in Fig. 20½, No. 7. The blade is 16x6 inches, square bottom, blade thin, light, and sharp, but curved cylindrically to strengthen it. It must be *thin* to cut well and sharpen itself; and *light*, to save lifting unnecessary weight with every spadeful. The handle is slightly bent, and has a T cross at the top. The kind I use are marked "Patent Ditching-spade, Antrim 2," and weigh 4 lbs. 2 oz. I bought mine of the George Worthington Co., Cleveland, O.

Second, the bottoming-spade, light and sharp, and curved like the other, 16x4 inches, but with cutting edge *rounding*, so as to leave a hollow groove a little over four inches wide in the bottom of the ditch for the tiles to lie in. This spade is seen in Fig. 20, next to the one first described, at the extreme right, and also in Fig. 20½, No. 6.

Third, the scoop for cleaning out the crumbs of earth left by the spade in the bottom of the first course. The blade is flat, and curved slightly, lengthwise (that is, across its length), so as to save friction in shoving it under the loose earth. It has a long handle, set at such an angle that the workman's back need not be much bent in using it. It is seen near the right of Fig. 20, its blade lying flat on the stoneboat, and its handle slanting up behind the handles of the two spades, and also in Fig. 20½, No. 4.

Fourth, the bottoming-scoop. This is shown in Fig. 20, a little to the left of the other, and in the same position, and in Fig. 20½, No. 1. It is a "double-ender," made of a half-cylinder of

FIG. 20½.

rather thin sheet steel, rounded and sharp at both ends, and hung near the middle by a ratchet (or clamp) arrangement that permits it to be adjusted to any angle. They are made of different sizes. I use a two-inch one (single-ended and not adjustable), and a four-inch one seen in the figure. The two-inch one is seen in the hands of the workman who stands in the ditch, Fig. 20, and is best for cutting a true groove for two-inch tiles after the other has cleaned out the crumbs, and left a wider-grooved bottom. The position of the workman in using the scoop is best shown in Fig. 26, page 94.

Fifth, the span-level, for determining grade. Mine is a home-made affair — simply a triangle, or Greek delta, 8¼ feet on each side, made of two-inch pine or poplar, plain battens, and has a spirit-level screwed accurately to the cross-batten that makes a capital letter A of the triangle. Care must be taken that the spirit-level be exactly parallel to the base of the triangle. For convenience in using, a short inch strip is tacked to each end of the bottom edge of the base. The level is graduated to show a grade of one, two, and three inches to the rod. Its use will be described later. It is seen in Fig. 20, standing on the stoneboat in the background, or, rather, back of the other tools.

Sixth, the tile-hook. It is seen in Fig. 20, in the hands of the first workman, who has a tile on it, which he is just in the act of laying in position in the ditch. It saves getting down into the ditch. But often I prefer to lay by hand, when the ditch is not too "nasty," standing on each tile as laid, and pressing it firmly into place.

Seventh, the filling-hook. It is seen leaning against the span-level, in Fig. 20; also in Fig. 20½, No. 8. Its tines are a foot long, and are flat, about ⅜ inch wide, and very strong. Mine is made by bending the shank of a potato-digging fork to a slightly acute angle, and setting it in a strong handle. You strike it with a sharp blow into the earth on the side of the ditch, and pull in 25 to 75 pounds at a time, according to how strong you feel. It is by far the best *filling* tool I know of, where one must fill by hand, as on turf or wheat, or in freezing weather. I also use the common potato-digging *fork*, seen at the right of Fig 20, behind the spades and scoop; also a shovel and hoe, and a strong garden-rake for cleaning up the crumbs on wheat, or on turf that is not to be plowed. These, I think, are all the tools really needed for hand-digging and for laying tiles, except the pick and crowbar for removing stones and loosening exceedingly solid gravel. Instruments for laying out the drains will be mentioned in connection with that work.

CHAPTER IX.

How to Drain; The Manipulations; Locating the Drains.

This has been described in part in Chapter VI. If there is plenty of fall, and the depressions show plainly where the main drains should be, the system may be laid out by the eye, with measuring-tape and stakes. If not, then a careful observation should be made *in very wet weather* (the fall and winter preceding), to locate the low parts, where the water gathers and flows off, and the direction the water takes on the more level parts. If the ground is very level you had better employ a really expert drainage engineer, or a civil engineer who understands tile drainage, to lay off the ground and set grade stakes, and make a diagram on paper. A diagram, or map, is an excellent thing to have at any rate. It is a permanent record of the location of drains. It helps one to find readily the main drains, if, at some future time, he wishes to find them to introduce new laterals. In my own case I have often desired to do this during the last 20 years, as some of my mains were laid even longer ago than that, and the laterals not all laid, in some cases, until many years later. The mains are thus easily found, and in every

TILE DRAINAGE. 85

case have been found where the map indicated, and to be as clean inside as if just washed out and rinsed out. That was just what was done at the last wet time preceding.

BEGINNING THE DIGGING.

Where shall we begin to dig and lay? At the outlet, as a rule. You *must* begin there if your work is to go on late in the fall, or all winter, or very early in the spring; that is, during freezing and thawing weather. You must dig and lay *and fill, as you go*. Sometimes the earth thrown out will begin to freeze in an hour or two enough to obstruct the filling, and, if left over night, it will often freeze so solid that it can not be filled in for days or even weeks. The sides of the ditch, too, will freeze if the ditch is left open long in winter; and when it thaws they will cave in or slump in. Still further, the earth fills in far more easily immediately after it is thrown out, and while it remains in spadefuls, than if left to settle together, especially if rain falls on it and soaks it into mud.

My plan for winter tiling (and it is just as good for other seasons if all is to be done by hand) is to lay and cover one entire main drain first, seeing to it that the outlet or outfall is good. Then begin digging at the lower end of the lower lateral, and dig right down to the main and take up one straight tile and insert a T or Y joint as the case may require. Of course, if the whole system is laid out *with the plow* in the fall, then the exact position of each lateral is fixed and clearly seen, and the T or Y joint may be laid at the time the main is laid, and its open part stopped with a flat stone. Either way will do. In the days when we could get no joints or junctions, T or Y, nor even get tiles with holes for lateral junctions cut in them before baking, we had to peck a hole in a tile, or at the junction of two tiles, with a sharp-pointed hammer, and that had to be done, or at least could be best done, when the main was first laid; and so I used to lay a few lengths of each lateral, as I came to each

86   TILE DRAINAGE.

in laying the main, and stop up the end of the lateral with a wad of straw until we began on each again. But with joints for laterals ready made, the first way is much the easiest. In either way the lateral *furrow* should be dammed up and the water turned off a few feet above where the lateral enters the main, so that surface-water will not flood down the furrow and wash out or choke up the main before the lateral is laid.

THE DIGGING.

"Any fool can dig?" No, it takes skill, won only by thought and practice, to dig rapidly and well, and not make hard work of it. I have had many men stronger than I, and with more power of endurance; but only one, I think, who could dig more rods of ditch in a day than I, and leave a true grade in the bottom. He was an expert ditcher who had followed the work as a business for years, and had splendid muscle, great endurance, and a true eye and hand.

DON'T BURY THE SPADE.

The first point in rapid, easy ditching, is to keep one side-edge of the spade *out of the earth, in sight*, each spadeful. Fig. 21 shows how a non-expert will bury both edges of the spade at *gh, ij, kl, mn*, and have harder work thereby, both in sinking the spade and in breaking off the slice of earth.

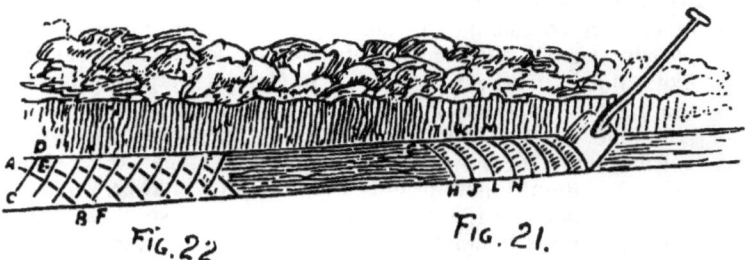

Fig. 22.—Expert digging. See printed page herewith.

Fig. 21.—Non-expert digging. See printed page herewith.

Fig. 22 shows how an expert will sink his spade; *ab, cd, ef*, being the curves cut by the spade, and the edge *a* being "out" the first cut, and the edge *c* being out the second cut, and so on. Thus held, the spade sinks more easily, and the *one* edge of the spadeful breaks off true and easily.

SINKING THE SPADE.

This is done by a succession of quick "shoves" or thrusts with the foot, throwing one's whole weight upon the spade with a quick impulse, and working the handle slightly back and forth in sympathy with the efforts of the foot. It will take from two or three to six or eight "shoves," to send a sixteen-inch spade "home," the number varying with the hardness or stoniness of the clay and the skill and muscle of the digger. The best way to get this motion is to watch a real expert and get him to teach you. I almost never use a pick or mattock. A good ditching spade well handled will dig almost any thing but the stoniest clayey gravel faster alone than with the help (?) of the pick, by working *around* the stones.

PRESERVING THE GRADE.

This is very important. I would rather dig and grade the bottom course complete than simply to grade an uneven "billowy" bottom dug by a careless, non-expert digger. If the proper grade is established at the bottom of the plowed furrows, as described, an expert will give exactly the same grade in the bottom of each hand-dug course, or "lift," simply by holding his spade, in digging, always *at the same angle* and thrusting it to the same depth, usually full depth. This, too, is a matter of practice, and requires skill to do it well. A green hand will always "shade out" where the ground is very hard, and "shade in" where it is soft; and he will unconsciously vary the angle at which he holds the spade. But the more nearly perpendicular it is held, the greater perpendicular or actual depth it will make when sunk full length. Exactly perpendicular, a sixteen-inch

88   TILE DRAINAGE.

spade will, of course, make 16 inches of perpendicular depth. Exactly horizontal, it will make no depth at all. Held at "half pitch," that is, at an angle of 45 degrees, it will make about 11¼ inches of perpendicular depth. As ordinarily held by a good ditcher, at a slight angle from a perpendicular, say 20 degrees, it will make about 14 inches. If the entire ditch is to be 30 inches, I usually try to make fully 7 or 8 with the plow, and 13 or 14 with the first spade, and that leaves only 8 or 9 inches for the second, or bottoming spade. The subsoil at the bottom is far more compact and hard, and it is better not to have too deep a course to dig. Great care should be taken to keep the grade of this course exactly right, so that, when you draw the double-ended crumb-cleaner and groove-cutter through the few loose crumbs of clay that are always left by the spade of even an expert, you will leave a true groove ready for the tiles. When the ground is very wet, or the atmosphere either moist or frosty, in digging and cleaning the bottom course I always keep the crumb cleaner, or "bottomer," close at hand, and clean out the bottom and "true up" the groove every 6 or 8 feet of digging, standing *at the top* of the bottom course, and thus never getting into the bottom of the ditch. The tiles, too, if the ditch is very muddy, are laid with the tile-hook shown in the hands of the first workman in Fig. 20. This saves one from getting very muddy.

THE FOOT-IRON FOR DIGGING.

One important little thing I forgot to describe among the digging-tools, and that is the foot-iron—see Fig. 23. It is made of plate iron or steel, about an eighth of an inch thick. It is fitted to the bottom of the hollow of the boot, between the heel and the ball of the foot. It is bent *down* behind about ¼ inch at right angles, to protect the boot-heel, and bent *up* on a curve about an inch along each side to protect the sides of the boot from wear against the handle and corners of the spade in digging. It is buckled over the instep

TILE DRAINAGE.

like a skate, and one strap passes around behind the heel, and is riveted or sewed, on each side, to the strap that passes over the instep. It should be fitted to the boot you expect

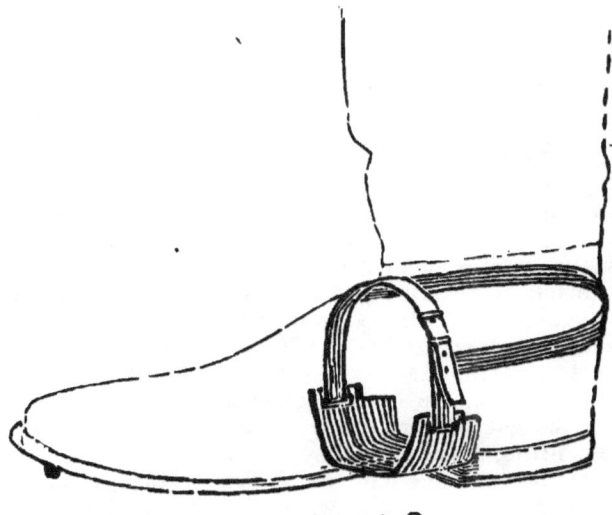

Fig. 23.—Steel "foot-iron," for shoving the spade (showing outline of boot-foot and leg, and iron buckled on for work.

to wear in ditching, by a skillful blacksmith and harness-maker, and should *fit just right*, both the iron and the straps. It should *buckle* on the *outside* of the foot, so that the buckle and loop shall not be in the way of the spade-handle. It is almost an absolute necessity with a rubber boot; and even a heavy leather sole is soon worn out without it, and almost immediately softens up so that it is spongy and not hard as it should be to thrust the spade most effectively. You can send the spade "home" far more quickly and easily with the ditching-iron, and it tires the foot less, and saves the boot. It is unbuckled, removed, and cleaned each night and noon. They should be kept ready made and of various sizes at hardware stores among ditching tools, but they are not.

PLACING THE EARTH IN DIGGING.

An expert will place the earth in a high, narrow ridge, *very close to the edge of the ditch.* Thus left it takes very little time or strength to fill it in, if it is filled almost as soon as dug.

THE MOTION IN THROWING OUT EARTH.

If your spade is held nearly vertical in digging, and the earth is either quite moist, so as to slip on the spade, or so dry as to crumble into small lumps, a part or the whole of the spadeful is inclined to escape from the smooth spade and remain in the ditch. A green hand will leave three or four times as much loose earth in the bottom of each "lift," of course, as an expert digger will. The expert first sinks his spade to the true grade, keeping one side-edge out as described, then pushes his spade-handle slightly forward to loosen the cut at the top, and then pries back with the handle, lifting at the same time with the lower hand, and, as soon as the side-edge and the bottom of the spadeful are loosened, gives the whole a swing, or curved motion, outward and upward, and lands *the whole of it*, spade on top, close to the edge of the ditch; and the spade is sinking for the next spadeful in a moment. The various motions described follow each other so rapidly that you can hardly distinguish them. The point is, that the spade *is swung with just such speed and motion,* and in such a curve, that the centrifugal force holds the earth against the spade-blade and prevents its either slipping or crumbling off. The speed and the curve must be regulated according to the tendency of the earth to slip or crumble. If it sticks to the spade tightly you can lift it as you choose. If very slippery or crumby you must give quite a quick swing, and keep it up until you land it. I remember that, when I first learned, as a child, to swing half a pailful

TILE DRAINAGE.

of milk over my head, bottom side up, aided by centrifugal force, I once unconsciously stopped, with pail in mid-air, to explain to admiring village friends just how the thing was done, and the milk descended ingloriously upon my foolish head! Just so, by not getting the right motion or "turn of the wrist," the green or inattentive hand lands much of his loose earth in the bottom of the ditch. To dig a tile drain rapidly, easily, well, and with true grade, is a trade of skill, and takes as much practice and knack as to learn how to use plane, saw, hammer, and chisel deftly and well. No, "any fool" can not dig tile drains and lay tiles well.

THE THREE-TINED DITCHING-SPADE.

In most of the prairie soils that require or need tiling, the opposite of slipping or crumbling takes place. The black, mucky earth is intensely sticky. It slices easily, being soft when moist, and being practically stoneless; but it sticks all over the spade, especially to its back, and will not "let go." You can't dump your spadeful clean, and must clean the spade every few spadefuls with trowel or large knife before it will slice the mucky earth again. Hence there has been invented, and, of course, patented, what we may call the three-tined

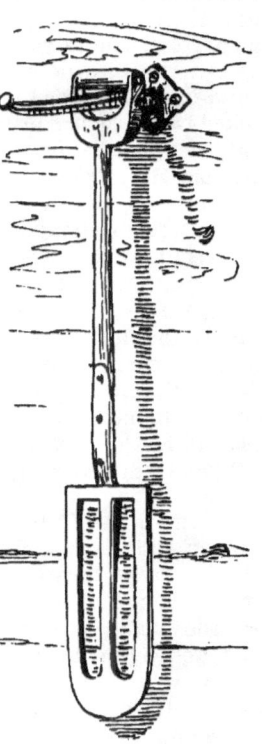

Fig. 24.—Patent three-tined ditching "spade," for very sticky and mucky soils.

ditching-spade. It is shown in Fig. 24, and consists of the ordinary spade-handle, but, instead of a full spade-blade, three strong, straight, steel tines extend down some 16 in., and are smoothly riveted to a spade-edge; *i. e.*, a cutting edge of about five or six inches wide and two inches deep. This slices the mucky earth nicely, while the latter can not stick to the narrow, square tapering tines. But these in turn help lift out the slice of earth. Such tined ditching-spades are found in most of the hardware stores of the prairie States. I looked in all the hardware stores of Cleveland, O., in vain to find one to try in my clay subsoil. It is so stony, and apt to crumble, that I presume the regular ditching-spade will work best here. I doubt whether the tined spade would take such dirt out clean enough; but I should have liked to try one, and wish now that I had sent west for one by express. But I did not, and hence it is simply my opinion that, on the average, in our clayey subsoils, the regular ditching-spade will do the best and cleanest work. Still, in some of our clays at a certain stage of dampness the earth sticks to the back of the spade quite badly, and so I usually carry an old table-knife in the hip-pocket or side leg-pocket of my overalls. It is the best and quickest thing to clean the spade with; and, thus carried, it is always ready for use.

ESTABLISHING CLOSE GRADES.

When there is an abundance of fall, as on most of our rolling clayey farms, a good hand and eye will *keep* the grade, when once established by the plow and flood-water, near the surface as described. But where the grade is pretty close, and you have only an inch or so to the hundred feet, it is best, as before stated, to get a careful civil or drainage engineer to establish the grade and set exact stakes as often as each hundred feet, each marked with the exact depth of the ditch at that point. Then you can yourself set *sighting-rods*, or *boning-rods*, as they are sometimes called; that is, set a stake about four feet high, about a foot from each edge of

TILE DRAINAGE.

the proposed ditch, at each hundred-foot engineer's stake, and tack a lath or batten exactly level across the ditch to the top of these stakes, so that the top of each lath, or sighting-rod, shall be exactly 66 inches (5½ feet), for example, above where the bottom of the ditch must be at that point, as shown by the engineer's figures. Then have a "sighting-stick," or batten with cross-stick at hand, itself also just 66 inches long (a convenient length for sighting over for the average man), and, as you dig your bottom course, occasionally set the sighting-stick *as nearly plumb as you can* (this is important) in the bottom of the ditch as there dug, and hold it plumb with one hand, and sight over it, thus held, to the next two or three sighting-rods down or up the ditch. If the top of your sighting-stick is in exact line with the sighting-rods, your ditch at that point is at the right depth. If it is not so, then make it so. Or, you may use your body as a sighting-stick,

Fig. 25.—Position of workman sighting over the "sighting stick" to the "sighting-rods." Here he stands in the ditch (one side represented as removed at that point), and sights along the ditch already dug.

setting your sighting-rods the exact height of your eye as you stand erect, and simply standing erect in the bottom of the ditch each time you sight for grade. Fig. 25 gives the sighting-rods and stakes, and the position of the workman in using the sighting-stick.

CUTTING THE GROOVES FOR THE TILES.

The position of the workman in doing the work is shown in Fig. 20 (page 80), the middle workman there being shown at that work. But it is shown more accurately in Fig. 26, which

Fig. 26.—Position of workman using the "bottoming-scoop" or "groove-cutter." Here he stands in the ditch, one side removed.

shows simply one plumb side of the ditch and a workman using the groove-scoop. My own custom is to remove the crumbs with the double-ended four-inch scoop, and then cut the groove for two or three inch tiles with a scoop of exactly the right size, so that the tiles (two or three inch laterals) shall lie snugly in the groove, unable to roll, and with no chance for water to flow *under* the tiles and gully the bottom out. In fall and winter the soil-water will commonly ooze into the groove enough to show whether there are any depressions in it, and whether the grade is right. *There should not be any depressions.* If there are any, then, after the tiles are laid, sediment may slowly lodge in them and fill up the tile partly or entirely. If there is not enough soil-water to

TILE DRAINAGE. 95

show grade, then the span-level should be carefully applied *to each eight feet of the groove* where there is any possible doubt; or, water may even be brought in small quantities and poured into a stretch of the groove before the tiles are laid. If it runs off properly then, it will do so ever afterward. I think you feel a little more *certain* you are right when the groove is actually tested with water than with either the sighting-rods or span-level, or both. But it saves time to use these first, and finally use the water.

LAYING THE TILES.

Where the sides, bottom, and top of the ditch are not too very muddy I prefer to stand *in the ditch* and lay each tile by hand, making sure that it will not roll or rock, forcing it up close to those already laid, with a slight backward swing of the boot-heel, and with your whole weight pressing each tile firmly into its groove and final resting-place. If the ditch is very muddy, the tile-hook (Fig. 20) may be used. This work should last for centuries if well done. It will not work well two years if badly done. It pays to do it *well*. Doctors are (slanderously?) said to "bury their worst mistakes." The owner of a farm can not afford to do the same in tile draining.

COVERING THE TILES.

I usually (unless I have a real expert working for me) lay the tiles myself, and have laid most of the 80,000 on my farm. I also prefer to fill the first course of earth myself, using fine damp clay (not big chunks), even if I have to slice down the damp ditch-sides to get such clay. I thus fill in about six inches and *tramp it hard with the feet*. This forces the water to seek the tiles through the natural pores of the undug sides of the ditch, and *to enter at the sides and bottom*, instead of washing large holes down through the loose soil, and filling up the drains. After the bottom course is thus filled and tramped, the rest may be filled in by man or team, and heaped up along the line, and trusted to settle with the rains of winter and spring, and rolled down with a heavy iron roller in the spring, or plowed and harrowed with the rest of the field if that is done.

THE UPPER ENDS OF LATERALS.

These should each be stopped up with a small flat stone, completely shutting out the dirt and the undue entrance of water. Water should enter there just as much as at other joints, and no more — that is, enter at the narrow crack.

PROTECTING THE OUTLETS.

It is well, where water or land vermin abound, to protect the outlets from their entrance by strong galvanized-iron screenwork. The following poetic (?) squib illustrates the importance of such screens. It appeared in *The Country Gentleman*, Jan. 22, 1880; and since that sedate periodical countenanced it then, we may, perhaps, venture to reprint it now. It was suggested by finding the bones of a rat washed out at a tile outlet.

THE DOLEFUL TALE OF A RAT.

Once a rat, in a rain,
Ran into a drain,
And said: "It is perfectly clear, sir,
It may thunder and pour
Outside of the door,
But it never can storm in here, sir.

" I'm ' as snug as a mouse '
In this fine dry house,
And fixed ' just as nice as a pin,' sir;
For so tight is the wall,
And the ceiling and all,
That the water can never work in, sir.

" And the door is so small,
And so narrow the hall,
That there's not the least fear of the cat, sir;
And so round is it dug,
And so tidy and snug,
'Twas just made for the home of a rat, sir ! "

TILE DRAINAGE.

So on ran the rat
Through the "main drain," for that
Was so straight, and so airy and big, sir.
He could gallop or trot,
Or lie down or "what not,"
Or stand up and dance you jig, sir.

"Ho, ho!" said the rat,
"Here's a smaller drain that
Was made for a bedroom, I know sir;
The door is quite small,
But I know how to crawl,
So into this bedroom I'll go, sir."

'Twas a pretty close fit,
But he squeezed into it,
And crept far along for to see, sir,
Whether, on as he went,
Along up the ascent,
Any bigger this bedroom would be, sir.

Then he lay down to dreams;
But the rain, as it seems,
Grew harder the longer he slept, sir,
And soaked through the "sile,"
And entered the tile,
And into his sleeping-room crept, sir.

"Ho, ho!" said the rat,
"What is that—what is that?
The waters?—the waters, they flow, sir!
I'll turn me about,
And 'skedaddle' right out,
For sure it is high time to go, sir!"

But, alas and alack!
He could neither go back
Nor forward, nor could he breathe there, sir,
For the waters they rose
Right over his nose,
And cut off his supply of fresh air, sir!

(If the reason why
When the rat was dry
He could enter the tile, you ask, sir,
And could not turn about
When wet, nor *back out*,
But was tight as the bung in a cask, sir—

Why, 'tis perfectly plain,
And not hard to explain;
For, you know—if you don't, then you ought 'er—
Though a thing is quite small,
It *don't stay so at all*
After soaking awhile in the water.)

So, alas and alack !
He could neither go back
Nor forward, nor stay, as I said, sir,
But was " laid out flat,"
" As a drownded rat,"
In fact, he was—"*mortuus est,*" sir.

Long after, 'twould seem,
Were borne down by the stream,
And found by the farmer, you see, sir,
Just the bones and all that
Of this ill-fated rat—
"As dead as a door-nail" could be, sir.

HÆC FABULA DOCET—TO RATS :
You may learn from the fate
Of your mis'able mate
To keep out of the mouth of the drain, sirs;
For though it *seems* dry,
There are good reasons why
You had better stay *out in the rain, sirs!*

HÆC FABULA DOCET—TO FARMERS:
At the mouth of your tile
It has happened erewhile
Such "varmints" have ventured to go, sirs;
But it's not at all hard
The outlet to guard,
With a grating they can not go through, sirs!

W. I. C.

French says, page 183 of "Farm Drainage," "They (the vermin after entering at the outlet) persevere upward and onward till they come, in more senses than one, to an untimely end. Perhaps, stuck fast in a small pipe tile, they die a nightmare death; or, perhaps, overtaken by a shower, of the effect of which, in their ignorance of the scientific principles of drainage they had no conception, they are drowned before they have time for deliverance from the strait in

which they find themselves, and so are left, as the poet strikingly expresses it, ' to lie in cold *obstruction* and to rot.' "
Then he speaks of the "slimy things that creep with legs," and which " seem to imagine that drains are constructed for their special accommodation." When they die in the drains they are affected as " sighing and grief " affected Falstaff— it " blows them up like a bladder," and, like Samson, " they do more mischief in their death than in all their life together. They swell up and stop the water entirely, or partially dam it, so that the effect of the work is impaired."

" To prevent injuries from this source there should be at every outlet a grating or screen of cast iron or of copper wire, to prevent the intrusion of vermin." The simplest way is to cut a square piece of heavy galvanized-iron screen or window grating, a little larger than the diameter of the tile at the outlet you wish to guard. Tack it firmly to two small wooden stakes, or wire it to iron ones, and drive one stake down firmly on each side of the outlet, with the screen pressed up tight against the tile. Then it can be removed for cleaning if necessary, and replaced. It is also well to have as few actual outlets as possible—gathering the various laterals and sub-mains all into one great outlet, and guarding that properly. The open ditch at each outlet is a source of constant annoyance. Cattle tramp it; frogs, crawfish, and slimy things gather in it and seek entrance to the drain, and the mud works in and obstructs the tiles unless it is often cleaned away. The fewer outlets the better. My largest outlet, ten-inch, discharges the water from over 25 acres of tiled land, and takes the flood-water from two ponds, and these in turn take the flood-water from 40 acres or more of untiled land. How this water is handled will be explained in another chapter. The point now is, that this one outlet takes the water collected in the two ponds, and, by some six miles of laterals and sub-mains, there being not less than 65 separate laterals and 8 separate sub-mains all discharging through this one outlet. And this one outlet takes no more

100   TILE DRAINAGE.

expense to screen it and keep it in proper flow, summer and winter, and watch the open ditch into which it flows, than each one of the laterals and sub-mains would require if each had a separate outlet. Indeed, it requires less; for it has such force and volume of discharge that it clears its own way. A map in the next chapter illustrates this more fully.

CHAPTER X.

How to Drain ; Special Problems.

One problem is, how to handle large amounts of surface-water coming from land that lies higher up along the watershed or slope. Such a problem presented itself in connection with the thorough drainage of the field of 36 acres shown in Fig. 27. First, let us explain the handling of the surplus water coming from some 40 acres of land, chiefly pasture and meadow, and not tiled, and lying higher up than the field itself.

I had the two ponds, A and B, at the southwest side of the field (see Fig. 27). Beneath the dam of the larger is a two-inch iron pipe with globe-valve for discharge full size of the pipe. This discharges into a sewer-pipe catch-basin marked E, which connects by a four-inch tile drain, and this in turn discharges down the valley through the middle (six-inch) main (see map). The valve is opened in high water. Also the overflow from both ponds in high water enters the six-inch main through the twenty-inch catch-basin in the catch-pond on the southwest side of the field, and marked C. The catch-basin is of twenty-inch sewer-pipe, and the six-inch main enters it by a reversed "trap" shown in Fig. 28.

EXPLANATION OF MAP, FIG. 27.

The single straight lines are two-inch lateral tile-drains, and the double lines are main drains. Southeast of the

TILE DRAINAGE 101

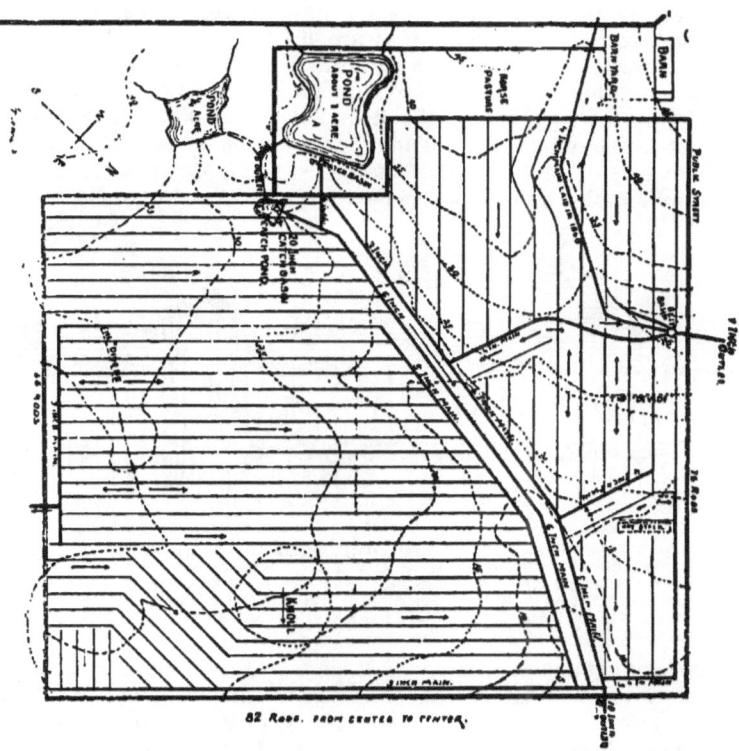

MAP OF "SPECIAL PROBLEM" ON 36 ACRES.

Fig. 27.—Thorough drainage of 36 acres (besides fence-rows and headlands), situated on the farm of W. I. Chamberlain. The curved and crooked dotted lines are lines of equal elevation above the outlet, "contour lines." They are marked 5, 10, 15, etc.; that is, the number of feet of rise or elevation above the outlet. The single straight lines are 2-inch lateral drains; the double lines are main drains. A is an ice and fish pond, and B a pond for watering stock. C is a "catch pond" (usually dry) in which is the "catch-water" or "catch-basin" to take the pond-water in freshets into the 6-inch main.

three large diagonal mains, the laterals are 33 feet apart, northwest of these mains they are 49½ feet apart. The former were laid eleven years ago, and drained the land so rapidly and thoroughly that, last fall and winter, when I drained the northwest part of the field I decided to put them 49½ feet (three rods) apart. Thus far they seem to be sufficient. The rule of running the laterals straight down the slope would make them cross all contour lines (the dotted curved

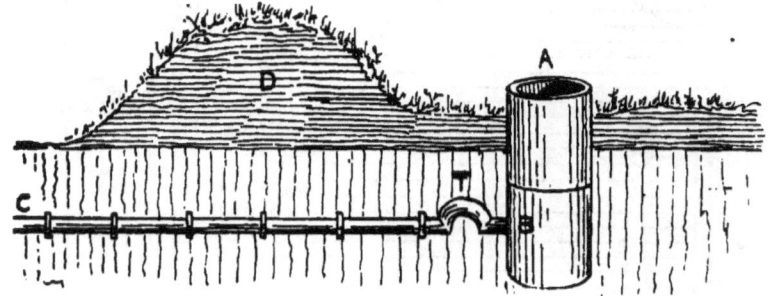

Fig. 28.—Sewer-pipe catch-basin, trap, 6-inch tiles, and section of dam (D) of catch-water pond (shown at C, Fig. 27). When the water overflows the two large ponds it rises in the catch-water pond until it overflows the sewer-pipe at A, and passes off through the trap T, and through the 6-inch main B C, and finally out at the 10-inch outlet O, near the northeast corner of the map, Fig. 27.

lines of equal elevation) *at right angles*. It will be seen that they do so as nearly as is possible if we are to have a parallel system, which is quite desirable, saving trouble and expense. Near the northeast side of the field a number of the laterals *deflect* and run due north several rods so as to avoid quite a knoll shown on the map. To run them straight northwest through this would have required deep and expensive digging.

A careful study of the "contour lines" and darts will show just what course is taken in the tiles by all the surplus water that falls on the entire field, and why the mains and

TILE DRAINAGE.

laterals were laid just where they are. The map will repay careful study.

SPECIAL PROBLEMS AND QUESTIONS.

I next touch upon several inquiries, problems, and statements that have come to me by mail.

From Sheboygan, Mich., came the following letter and diagram which I answered in *The National Stockman* in substance as below. I quote with due credit.

"I have a wet, boggy piece of land, too wet to pasture, which I wish to drain into two or three reservoirs. This diagram shows the shape, slope, and features of the land, and my plan concerning it. The darts show the slope. Could this piece be drained by cutting one ditch toward the southeast along the dotted line A B (see cut), another southwest as shown by the other dotted line

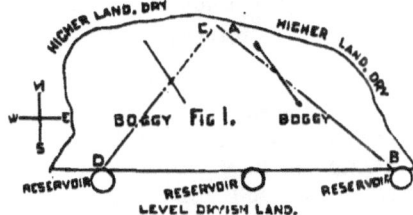

Fig. 29.—Problem submitted.

C D, or will it be necessary to cut other ditches running parallel to the north and west ditches [I don't understand this], or to run from A to S? Also how far apart, how deep, and what sized tiles? Should they be laid close, and be mortared on top, or a little way apart on top? Should the ditch have a good pitch?"

Now, this letter is too indefinite. The amount of fall is not given, nor the reason why he wants to drain into reservoirs. Assuming that he has reasons, that he can't get other (free) outlet, and that these reservoirs have gravel bottoms, and that they will dispose of the water from the bog so that it will not get back into the tiles, I answer the queries as follows:

Such a bog, unless the soil is quite porous, would need drains 30 to 50 feet apart and 2½ to 3½ feet deep. As to pitch, get all that the land will give you. If nearly a dead level, "shade up" a trifle, say from 3½ feet deep at the outlet, to 2½ feet deep at the upper end. As to size of tiles, I would use none less than three or four inch in a "bog." Lay the tiles as close as possible, and use no mortar. Simply cover with clay, if possible, six inches deep, and tramp well before filling the rest. If this is your first job of draining you should get either a good drainage engineer or an expert practical tile-ditcher to start you and show you how. From your inadequate description I should say the land should be laid out as below, in the rough sketch:

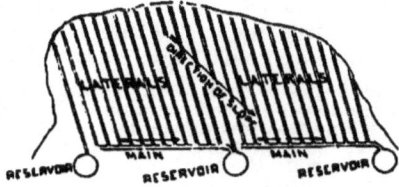

Fig. 30.—Solution suggested.

This is as near as I can hit the case from the insufficient data given above.

DRAINS WITH NO OPEN OUTLETS.

This raises the general question of drains without open outlets. All through the gravelly drift soils and subsoils of Ohio may be found cut-like depressions so deep that they can not be drained through to lower ground except by cutting 6, 10, and even 15 feet deep, or more, through intervening land, which would cost too much. These small pockets are often excellent land, only subject to flooding by sudden showers running down the adjacent slopes and not soaking away until the crop, say of wheat or potatoes, is perhaps ruined. If the adjacent subsoil is gravel, the problem may be solved so that it will pay. Mr. T. B. Terry, of Hudson, O., has drained a rather difficult one of this sort lately (not his first) by running a main drain right into the bank toward where the outlet would be if the main were carried far enough. This bank he found, as he expected, to have a porous gravel subsoil. Into this he ran the main, several rods, six feet deep or more (if I remember his statement correctly), until he judged the distance gave the gravel time and capacity to absorb from the tile main the water brought by it from the cuplike depression during heavy showers or snow-thaws. Then in the depression he laid a system of laterals, pretty close together, and joined each to the main, to take the surface-water down and into the main rapidly. He reports that it works perfectly, as do other smaller systems of drains without outlets, on his farm and others. But for this you must have a gravelly, sandy, or other quite porous subsoil.

SILT OR SAND WORKING IN AT THE JOINTS.

On this subject, as well as the foregoing, I have had no experience, as all my drainage has been where the subsoil is clayey. But I have received so valuable a letter from Mr. W. Trowbridge, of Painesville, O., that I here give the main part of it.

This and most of the other queries and suggestions that follow came to me because of my articles in *The Ohio Farmer*, and most of them have been answered in that paper in substance as here. Mr. Trowbridge says:

I do not now remember of your saying any thing about covering the joints of the tile with any other material than earth. Now, will you please allow me to make a suggestion or two?

1. As to covering the joints, perhaps in a tenacious clay no other covering than the material taken from the ditch is needed. But *soils requiring draining are not all clays*. Where I have done the most of my work, the soil is *sandy* and a *black friable loam*, underlaid at a depth of from two to five feet with clay. The question was, how to keep the fine sand from entering and filling the tile. I think I have solved the problem in a very satisfactory way, at least to myself. I am using common building-paper, cut into pieces to suit the size of the tile (for 2¼-inch tile, 2x8¼ inches is about right). One of these pieces, laid over each joint, effectually prevents the silt from entering the tile, at least as far as the paper extends. In laying the tile I always carry a mason's trowel in my hand, and, after placing the two ends of the tile as closely together as possible, place a trowelful of earth directly on each paper as laid.

I have uncovered tile after it has been in the ground three years, and always found the paper intact, and excluding the dirt as well as when first laid, but, of course, rotten, and of no account when disturbed.

This idea of using *paper* is not original with me. I had it from Col. Waring, in an article of his published (not in his book) in (I think) the *American Agriculturist*, years after his book was issued. But the use of *building-paper* is an idea of my own. I buy it by the roll, expressly for this purpose.

ESTABLISHING GRADE.

On this point the same letter says:

Another point, which I think to be very essential, is to get the bottom of the ditch of a true and uniform grade, especially where there is but a slight fall For this purpose we use a light, strong linen line stretched directly over and at a certain height above the proposed bottom of the ditch (7 feet is a convenient height); then with a light wooden rod 7 feet long the workman can cut down so the rod will just set under the line. By using the rod at frequent intervals, a careful workman can, with absolute certainty, make a true and even grade. The line must be supported at intervals of about 60 feet.

This letter is valuable because written out of *actual, successful experience*. The line he recommends is used instead of the "boning-rods," mentioned before in this little book, and recommended by Waring and others.

QUICKSAND POCKETS.

I have, in a few places, struck small quicksand pockets just where the tile should lie. To prevent the tiles from sinking out of true I have laid strips of board, 1x4 or 1x6 inches, and of the necessary length, right on the quicksand at the proper grade, and laid the tiles on these, and covered with clay, with strips of tin or of heavy paper, over the joints. Tarred or oiled building-paper would be best. The work has been successful. Others who have had longer strips of quicksand have reported success by the same means.

SILT-BASINS, "MAN-HOLES," ETC.

I have never found these necessary for my soil, with my excellent fall. The Storrs & Harrison Co. (nurseries) Painesville, O., have tiled large areas of very level land near Lake

Erie, in Painesville. Silt, or fine sediment, troubles them considerably. At intervals they simply sink shallow wells in the course of long laterals, some three feet deeper than the drain, and stone them up, running the laterals into and out of these wells. The silt settles in these wells, or "silt-basins," and can be cleaned out in dry times. I understand, too, that they grade mainly by soil-water, digging when the ground is wet and nursery work not pressing, and that they open up an entire stretch of lateral and begin to lay tiles from the upper end, so that the muddy soil water (muddier while digging) may not run into the tiles from above, and partly choke them in construction. Many of their laterals, too, empty into deep open ditches, cleaned out as often as necessary, and bridged as required. In such cases the last few tiles near the outlet are laid on a narrow board, which projects into the open ditch a foot or so, and helps keep the outlet clear and unobstructed.

USE OF STONES.

Mr. A. B. Cowan, of Morgan Co., says:

I have land that "slips," and is in permanent pasture. I don't intend to plow it, as it is heavy limestone soil, and well set in blue-grass. The question is, how to drain it, as it is rough. There is a stone-quarry above, and tons of cobble-stones. Would you advise digging a ditch and then break them and throw them in? Or would it be better to build in the stone so as to make a passageway? I have a quarry where I can get stone out from $\frac{1}{4}$ to 2 inches thick, and as smooth as slate.

I would use tiles rather than stones. They require less width of ditch, and hence the expense of digging is less. They are laid much more rapidly than stones, and last longer unobstructed. But if you decide to use stones, then lay a row of long and rather square cobbles on each side of the bottom of the ditch, and cover with flat stones, two courses deep, and then throw in a foot deep of loose stones, and then fill with soil. This is the way I laid drains for my father 40 years ago, before tiles were made in this region, and they worked well for many years. This plan makes a regular channel. Stone drains do pretty well if each has its outlet, but you can not well lay out a system of mains and laterals, as with tiles. I bury hundreds of loads of stones in excavations in sidehills, and grade and plow over them to get them out of the way, and use only tiles for drains.

ROOTS IN TILE DRAINS.

W. B. Hall, of Strambury, Pa., asks about proper *depth*, *distance apart*, and whether roots from a vineyard will obstruct the drains. Depth, 2½ to 3½ feet. Distance, 2 to 3 rods in compact clay; further in more porous soils. The question of roots in drains is an important one, and I can not do better than to quote from an article by myself, in the *Country Gentleman* for June 17, 1880, with such notes as later experience seems to require.

IS THERE DANGER FROM ROOTS?

Yes, and no. No, under ordinary circumstances; yes, under occasional conditions. It does not seem to be, on the part of the roots, a question of ability, but of desire. They can enter the tiles without the least difficulty, but they do not usually desire to do so—at least, so as to obstruct the drains.

As to the *ability* of the roots, even of wheat and clover, and other cereals and grasses, to enter the drains, I have not now the slightest doubt. I formerly supposed the cases given in the books, of wheat roots, etc., growing to a depth of three feet, and even four, were somewhat exceptional, or applied only to mellow subsoil, and that in a very hard, stiff clay they would not extend so deep. But I have lately finished laying about three miles of tiles, three feet deep, in a wheat-field where wheat grew last year too; and every spadeful that I took the trouble to examine, even from the bottom course, had very many fine, live wheat roots all through it, and many nearly decayed roots of last year's wheat. How much deeper than three feet they went I did not dig to ascertain, except that our drains sometimes, from unevenness of surface, went as low as three and a half feet, but never below the roots. Though not in my present line of thought, I can not but notice the wisdom of this provision of nature. The surface was very dry—no rain for four weeks; but the subsoil at the depth of three feet, or even two, was quite moist, and these roots pumped water to keep the wheat from withering. Wheat, we all know, stands drouth wonderfully, especially on clayey loam, and the reason is plain.

WHY ROOTS DO NOT USUALLY OBSTRUCT DRAINS.

Because they do not wish to do so. In a well-constructed drain there is ordinarily no inducement. A few days ago I took up a joint of an old main to insert a new lateral, and examined closely. There were plenty of wheat roots about the tile, and a few had feebly entered the cracks. But finding no moisture and no soil or "silt"—in short, nothing but a dry, empty hole—they had given it up as a bad job, and literally backed out and gone again into the soil for moisture. They concluded they couldn't "make" any thing there, according to the doctrine, "*Ex nihilo nihil fit*"— "You can't make something out of nothing." And in well-constructed drains in clay soil, or in any soil that is not "springy," this will, I think, be the ordinary result. At all events, the drains on such soils *do last* many years, with no tendency to stoppage from roots. There is no moisture in the drains to lure the roots, except when there is too much moisture everywhere else.

When will roots obstruct drains? I answer: When there is water or damp silt in the drains and dryness in the soil. This will occur in improperly constructed drains, or when there is perennial water in the drains from springs or springy ground higher up the slope, and dryness in the soil through which the drains pass, and on which are growing crops. Even in ordinarily dry soils, with no running water in the tiles in summer, if there are "dips," or depressions, in the drain, water will stand confined for some time after the flow ceases, and silt will lodge, and roots will enter and help make the stoppage complete and permanent. The remedy—or, rather, preventive—here is to construct the drain on a true grade, and have no "dips." In case of perennial water in the tiles, the remedy is not so simple. If the soil through which the water flows becomes itself dry, the roots will surely seek the water in the tiles. How shall they be excluded, and yet the water be admitted from the adjacent soil, as well as from the spring or springy ground above? The only way I know is to use soft, porous tiles, exclude all that have "pin-holes" in them, and lay the joints in hydraulic cement. The water will enter the pores of the tiles, but the roots can not. If any one doubts about the water entering, let him lay out a few large, soft-burnt tiles in a commencing shower, and see how they swallow the big drops as soon as the latter fall. Or let him set such a tile on end in plaster of Paris, or cement, on a board or plate; and when the cement or plaster is hardened, fill the tile full of water. But it must be quite soft-baked and porous, and such tiles are not so durable, and should never, I think, be used except in case of such danger from roots.*

---

* NOTE.—*Sept. 18, 1891.*—This was written in 1880. I had not then (I may as well confess) *actually made the experiment* of setting a soft-baked tile on its end in plaster of Paris and filling it with water, but I took

IS THERE DANGER FROM TREE-ROOTS?

Yes, great danger under the same conditions as in the case of cereals. The roots of aquatic, or water-loving trees, like the willow and some kinds of elm, seem to have almost no limit to their growth, either horizontally or vertically; and they seem to go in search of moisture or richness as if by instinct, and to know just where to find it. I have traced the roots of a smallish elm some 25 feet horizontally, and 6 feet vertically, to their feeding-place in a grave in an old cemetery; and I have, in plowing, traced the roots of a large elm *one hundred feet* horizontally by measurement. These roots will enter even a "pin-hole" in tiles, if they can find running water. Mr. H. B. Camp, of Cuyahoga Falls, O., told me a few days since that he once helped to take up an obstructed drain whose joints were laid in cement, I think he said. At all events, a willow root had entered at a pin-hole not larger than a small darning-needle, and spread into a fibrous mass and *packed the tile full* of roots for several feet — the only connection with the upper world and their lungs (the leaves of the tree) being this small, threadlike root that entered at the pin-hole. It is better to cut such trees down when they are near a damp drain, *and see that they are dead*. Prof. Townshend, of the Ohio Agricultural College, Columbus, exhibited in one of his lectures last winter a dark, stiff, fibrous, spongelike, solid cylinder, some three feet long and three inches in diameter. When the class had done guessing, he gave its history. It was the *willow-root core of his cellar-drain*. Knowing the nature of these roots he cut the willow down before he laid the drain, and burned the stump all he could; but in spite of that, its roots stopped his cellar-drain at a depth of five or six feet.

My experience and observation lead me to these conclusions:

1. In ordinary cases there is no danger from roots; and very hard-burned and even glazed tiles should be used, the water entering at the joints. Hard tiles are more durable.

2. Where there *is* danger from roots, as in the cases described above, soft, porous tiles should be used, with joints laid in cement. They may not last so long, but they seem to be the only kind that will admit water and exclude roots. [See the note above.]

3. Mere "collars" without cement will not exclude roots altogether, and great care must be taken to use no tiles that have pin-holes in them.

4. A uniform grade is important, or at least there must be no "up grade" between the head and the outlet. It will probably cause permanent stoppage.

---

the word of writers who claimed to have done so. Very thorough experiments since made, and reported in Chapter VII. (pages 7—76), seemed to prove that water will not enter a drain made as advised, viz., with soft-baked tiles and cemented joints, fast enough to amount to any thing. Of this, again at the close of the quotation.

5. All aquatic trees near the line of the drain should be killed before the drain is laid. They are a pest.

And I notice that English writers on drainage have been led by similar experience and observation to similar conclusions.

W. I. CHAMBERLAIN.

On the above I now (1891) remark, that, where water from a spring or springy place is to be *conveyed through* soil that is dry or likely to be so in summer, I would convey it in hard tiles with cemented joints, and drain the land locally in the ordinary way. As to the desire of roots to enter drains in common clayey soil, I may say that I have lately dug down to my drains in an apple-orchard set 21 years ago, with trees 33 feet apart, and whose roots long since met and passed each other, and whose branches have nearly met; and the drains are wholly free from any obstruction by roots.

But slop-drains, bath-room, water-closet, and sink-drains from the house, which are *constantly damp and often flushed*, and which often run past or near evergreens and various sorts of shrubbery around the house, such drains should be of glazed sewer-pipe, socket joints, laid in water-lime cement, or the roots, especially of evergreens and aquatic trees, will obstruct them.

DRAINAGE BY STEAM-PUMPING.

Chas. S. Killmer, Arenac, Mich., writes:

There are immense tracts of low-lying lands bordering inland lakes which I believe may be reclaimed by artificial means, viz., steam-pumps or wind-engines. I understand the question has been favorably decided in several instances near Bay City, Mich., and would refer you to the Wilson Hoop Co., West Bay City, Mich., for information. They have, I hear, a large tract of land successfully drained by this process. It seems to me that, in view of records in Holland and other countries, our leading men have strangely neglected a very important matter; and while our thickly populated States are breeding miasma from their countless marshes, our statesmen (!) are asking for millions to develop the desert regions by irrigation for the benefit of railroad corporations.

I give this suggestive letter nearly entire, so that any who choose may investigate. It did not seem to me to lie quite within the scope of this little book to investigate and develop the matter. Its scope is to show advantages and methods of drainage for clayey and swampy lands that have outlets, without expensive pumping. There are many million acres of such land ; and until much more of this is tiled, and land advances in price more nearly to prices in Holland, it does not seem to me that drainage by pumping will pay the individual farmer unless it be in exceptional cases. When the time comes that large areas of such land should be drained it will probably need to be done by taxation placed upon the land reclaimed, and proportioned to benefits to the land.

CHAPTER XI.

Estimates: Sizes of Tiles Required: Cost of Draining if Done Economically.

SIZES OF TILE REQUIRED.

I discussed this matter quite fully in *The Country Gentleman* of Jan. 9, 1879. I have since seen no reason to change my views, and therefore give that carefully prepared article nearly entire, with due credit to that excellent paper:

THE SIZE OF DRAINS.

About the size of drains, I can not agree with Engineer and P. Q., either theoretically or practically. We can never figure it out from the *annual* rainfall and the discharging capacity of a given-sized tile at a given rate, as, indeed, Engineer intimates. But if we try, we must at least figure carefully on our assumed data. For example, when we are told (page 759) that a one-inch tile, discharging at the rate of four miles per hour, will discharge all the water that falls on 36 acres of land, we begin to figure as requested. The rainfall for Ohio and most of New York we find to be from 32 to 44 inches—average 38 inches. In parts of Eastern New York and of New England it is 44 to 56 inches—average 50. [*American Cyclopædia*, last edition.] Take 38 as the basis. This gives about 23¾ gallons to the square foot, from which we easily find the number of gallons on 36 acres. By use of a formula explained in geometry, and constantly used in trigonometry and engineering, we get the area of a cross-section of the inch tile, and from this and the rate of flow we readily find the number of gallons discharged each hour, which is about 862. Dividing the whole number of gallons by this, and reducing, we find it would take over 4 years and 11 months to discharge the rainfall of one year, even if the tile works up to full capacity steadily all the time. And even here the rate is overestimated, for none but a very heavy grade will give a rate of four miles an hour in an inch tile, owing to the great loss by friction in small rough tubes.

Now, all such calculations are utterly unreliable, however accurately made, unless they are based on careful experiment, and take into account all the variations of friction in tiles of different caliber and inside smoothness, and varying pressure caused by different grades, etc. They mislead in both directions. On the one hand they assume that the tiles must discharge all the water that falls on a field, whereas they discharge only the surplus beyond the point of saturation, and sometimes for six or eight months evaporation and absorption by the growth of crops leave not a drop to reach the tiles. On the other hand, they suppose a uniform daily or weekly rainfall, just up to, and never beyond, the capacity of the tiles, whereas the tiles are sometimes idle, as we have seen, for months, and sometimes, on a soil already saturated, comes a rainfall of six inches in a single week, or even three inches in a single shower. Now, the tiles must convey all this away promptly, or it will wash and gully the surface-earth, or stand stagnant for days. We must calculate, then, for the maximum ever required of the tiles. To calculate according to the annual rainfall is like calculating the annual traffic to be borne up by a bridge, and from that estimating how strong it must be each minute; whereas we know that, as with the "wonderful one-hoss shay," so with the bridge, "the weakes' spot must stan' the strain," even the heaviest strain that can ever come upon it ; and the bridge must be known, from formulas and calculations based on experiment, to have as its "factor of safety" at least five times the strength ever likely to be required of it. The steel wire cables of the Brooklyn bridge have a strength of 80 tons to the square inch of section, and the four main cables are each to be 16 inches in diameter, so that the aggregate strength of the main span will be immensely beyond the combined force of wind, storm, and burdens ever transported.

These facts we all know. Now, it is the same way in drainage, though the risks involved are not so enormous. The main drain or drains must be up to the greatest emergency, or there is risk of partial or perhaps total failure. What, then, are the greatest emergencies? Facts alone can determine. Take a single one as a sample. This year (1878) here, while wheat stood in the shock we had over 3 inches of rain in 24 hours, on my farm. The ground was fully saturated by previous rains, and there was but little evaporation for a few days, and but little absorption by growth, as the wheat was cut. Therefore the drains had all this water to handle. Suppose, now (by P. Q.'s rule), we have one four-inch drain as an outlet for forty acres of laterals. Take Engineer's rate of flow, viz., four miles per hour, though it will require a pretty heavy grade and full pressure to give as high a rate. If I figure correctly it would take the main nearly ten days to free the field simply of surface-water; and unless there was surface drainage the shocks would have to stand in the water a week, as thousands of acres did on undrained land in the Black Swamp region near Toledo.

On my own farm one field (where there was wheat) happened to have 3 four-inch mains to take the water gathered by 12 acres of laterals. The "lay of the field" required three separate mains, and I was determined, from previous experience, to have them large enough, and so used four-inch tile for the lower half of each. They are none too large. Not less than five times within a year they have been crowded to their utmost capacity, even with surface drainage for a short time. In the case of the big rain mentioned, they freed the ground of surface-water in one day — that is, as fast as it fell, as they should according to the above estimate. Each had four instead of forty acres to drain, and did it in one day instead of ten. The average grade of the field is three feet to the hundred.

The rule given by the essayist at Hartford (mentioned by Engineer) is based on correct principles for a limited range of sizes of tiles and variations of grade. It is, in brief, "To find how many acres a given-sized main will drain, square its diameter." Thus, a three-inch main should drain nine acres; a four-inch one sixteen, and so on. But for our soil and variable rainfall I am sure this gives far too many acres. I should say, for sizes from three to six inches, and grades less than three feet to the hundred, square the diameter and divide by four. Thus:

A three-inch main will drain 2¼ acres.
A four-inch main will drain 4 acres.
A five-inch main will drain 6¼ acres.
A six-inch main will drain 9 acres.

For heavier grades it may do to divide by three. Thus:

A three-inch main will drain 3 acres.
A four-inch main will drain 5⅓ acres.
A five-inch main will drain 8⅓ acres.
A six-inch main will drain 12 acres.

But it must be borne in mind, that, the steeper the grade, the greater the danger of surface wash, which often causes great loss of manure, and even of soil itself.

My conclusions in brief would be: 1. Have your mains large enough — better too large than too small. Don't economize here. It will be "saving at the spigot and wasting at the bung."

2. By "large enough" I mean so large as to take the water as fast as the soil can filter it and the laterals collect it.

3. Expense may be saved by diminishing the size of the mains toward their upper ends. For example, three-inch will do till it has received water from two acres of laterals, then four-inch up to four acres, then five-inch up to six acres, and so on.

Since the above was written we have had twelve hours of moderate rain which has melted four inches of snow; total water, over an inch. The soil was saturated before ; surface not frozen. I have been to examine the main drains on my own and two neighboring farms. They are as follows:

Two two-inch tile drains draining about 1 acre each.
Two three-inch tile drains draining about 1½ acres each.
Seven four-inch tile drains draining about 4 acres each.
Five four-inch tile drains draining about 5¼ acres each.
One six-inch tile drain draining about 12 acres.

All are worked up to full capacity. The first three sets in the above table come under the first rule given above. They have a small surplus of water that they can not carry. The last two sets come under the second rule. They have quite a large surplus. In the case of a few, surface wash comes on from undrained land lying above. Due allowance is made for this.

W. I. CHAMBERLAIN.

I think our American drainage authorities recommend *sizes far too small.* The latest of these authorities, C. G. Elliott, a civil and drainage engineer whose work has been chiefly among the level soils and large rainfalls of Illinois, says : "For drains not less than three feet deep, and with grades of not less than 3 inches to the 100 feet, — for such drains not more than 500 feet long, a two-inch tile will drain 2 acres. Lines more than 500 feet long should not be laid of two-inch tiles."

Putting his rules in tabular form we have :

A 2-inch tile will drain 2 acres.
A 3-inch tile will drain 5 acres.
A 4-inch tile will drain 12 acres.
A 5-inch tile will drain 20 acres.
A 6-inch tile will drain 40 acres.
A 7-inch tile will drain 60 acres.

Waring says : " In view of all the information that can be gathered on the subject, the following directions are given as perfectly reliable for drains four feet or more in depth, laid on a well-regulated fall of even 3 inches in 100 feet :

For 2 acres, 1¼-inch pipes (with collars).
For 8 acres, 2¼-inch pipes (with collars).
For 20 acres, 3¼-inch pipes.
For 40 acres, two 3½-inch pipes, or one 5-inch sole tile.
For 50 acres, 6-inch pipes or sole tile.
For 100 acres, 8-inch pipes, or two 6-inch sole tiles.

"It is not intended that these drains will immediately remove all the water of the heaviest storm, but they will always remove it fast enough for practical purposes."

I do not think they will.

French gives several pages of dense and discouraging tabulated figures showing the velocity in feet per second and discharge in gallons per 24 hours for many sizes of tiles and grades of fall; but he gives, so far as I can find, no definite recommendations. He says : "The size of tiles is a matter of much importance. Tiles should be large enough to carry off in a reasonable time all the surplus water that may fall upon the land. Here the English rules will not be safe for us ; for although England has many more rainy days than we have, yet we have more inches of water from the clouds in a year. Instead of their eternal drizzle we have thunder-showers in summer, and in spring and autumn northeast storms, when the windows of heaven are opened, and a deluge, except in duration, bursts upon us. Snows cover the fields until April (in the North), when they suddenly dissolve, often under heavy showers of rain, and planting-time is at once upon us. It is desirable that all the snow and rain water should pass through the soil into the drains, instead of overflowing the surface, so as to save the elements of fertility with which such water abounds, and also to prevent the washing of the soil. We require, then, a greater capacity of drainage, and larger tiles, than do the English, for our drains must do a greater work than theirs, and in less time."

And yet he gives no tangible rule or opinion as to what sized mains are required here for various areas.

Klippart quotes French's tables, and, in substance, his remarks about the English drizzle, but gives no definite or tangible rule or formula.

TILE DRAINAGE. 117

From much observation I am convinced that all or nearly all the authorities down to date have advised too small tiles, basing their judgment more upon English and Continental practice than upon American conditions. The general estimates or rules given in my article quoted just above, from *The Country Gentleman*, are based upon much observation, and tested by many years of careful experience on my own farm ; and I believe they are safe for grades of less than 3 ft. per 100 ft (and, of course, for greater grades); and for annual rainfall not exceeding 50 inches, with occasional phenomenal rainfalls of 2 or even 4 inches in 24 hours. The sizes there given will carry the water as fast as the soil can filter it into the drains, if the latter are 30 inches deep—deep enough for clay.

The tendency toward larger tiles, especially in the rather level prairie West, is manifest and wise. The soil there is more porous, and hence laterals may be much further apart, and wisely laid deeper (even 4 or 4½ ft. deep) than in our more compact clayey soils in Ohio,(where 30 inches is as deep as best economy will warrant). Also, as the grades there are usually less, the sizes must be larger. The manufacture of 1 and 1½ inch tiles has long been discontinued, even in Ohio, and few 2-inch ones are now made in some sections, though they are *large enough* for an outlet for an acre, with good grade. But in Illinois, 3 and 4 inch tiles are now the smallest sizes found at most tile-kilns. The material is not expensive, and the tendency toward larger sizes is wise, except where freights or long hauling makes the weight important.

COST OF DRAINAGE IF DONE ECONOMICALLY.

In a series of articles in *The Country Gentleman* in 1878 I discussed the cost, and gave actual figures from a job of 15 acres, where much of the work was done with teams. The ground was previously plowed so as to leave a very deep dead-furrow just where each lateral was to be. In this a four-horse team (four abreast) plowed back and forth once with a heavy plow in a pretty damp time, making a total depth of some 15 or 16 inches. A single spade-depth with a long spade, and a groove cut with the bottoming-scoop, finished the digging, and most of the filling was done with the plow and team; and in the first plowing of the land thereafter the back furrows (as usual) were thrown into the former dead-furrows, and the filling was complete, and the land leveled up. The conclusion of No. 3 of this series, *The Country Gentleman*, April 11, 1878, is as follows, slightly condensed:

Two men can thus average fifty rods a day together (each taking a ditch thus plowed, and digging one depth in it). I follow; cut a true smooth groove about an inch and a half deep, with the regular ditchers' "scoop" or groove-cutter; distribute the tiles from the piles, lay them carefully *by hand*, walking on them as I lay them, to settle them firmly; and fill in the earth thrown out *by the spades*, packing it firmly around and above the tiles. Thus three of us do the entire handwork of fifty rods in a day. The rest of the clay and all the soil is to be filled in with team and plow after the handwork is all done. Each night, or whenever rain interrupts, all the handwork should be finished as far as begun—straw wads put in the exposed ends of the tiles, and the water turned aside from the ditch above, if necessary.

If the field is very stony, two men can hardly dig fifty rods this one course, but in most of the fields of mine and neighboring farms they can. I used to dig the bottom course with a very narrow "bottoming-spade," just wide enough to receive the tiles. But I find that most men will dig faster with a six-inch or a four-inch spade, and one can cut the gutter, or groove, better if the ditch is wide enough to walk in, and to curve the groove slightly to avoid fixed stones. In case of a very heavy stone, sunk directly in the line of the ditch, I find it better to curve the whole ditch gradually a foot or two right or left than to spend time to remove or sink the stone. The curve is easily made, as the upper part of the ditch is already made so wide with the team.*

In filling the rest of the ditch, the loose clay thrown out by the plow should first be plowed back, the off horse walking in the ditch to tramp it; then a furrow or two of soil from each side, the off horse still walking directly over the tiles as much of the time as possible. One day with man and team will do all the team-filling for the field.

We have, then, the entire cost of draining the fifteen acres 2½ feet deep, and 2 rods apart, taking present prices of tile and labor here:

---

* Such a curve is shown in **Fig. 20**, page 80.

TILE DRAINAGE.   119

```
20 ditches, 58 rods each = 1160 rods of 2 inch tile at 16 cts. per rod.$185.60
  ( 27 rods of 3-inch tile at 32 cts. per rod...........................   8.64
* { 27 rods of 4-inch tile at 48 cts. per rod..........................  12.96
  ( 44 days' work of common ditchers at $1.12½ per day...........  49.50
22 days' work of head ditcher at $1.50 per day......... .........   33.00
Cost of plowing ditch with four-horse team.......................   7.00
Cost of filling with two-horse team.................................   3.00
Cost of drawing tile from the yard or car, one mile distant......  10.00
Use of tools.......................................................   3.00
Allow for hindrances, interruptions, and extra stony spots (time).  10.00
                                                        Total..........................................$322.70
```
Dividing by 15 we have $21 51 as the entire cost per acre.

If the field is very uneven, so that the laterals can not be parallel with either of its sides, or even with each other, or so as to require deep digging in places, or a greater number of main drains and junctions, the cost will be increased. Still, the plow can be used to great advantage even then by simply turning out one furrow or two each way and sinking a third as described, wherever a drain is to run. In this case the four-horse team should be used all the time. I know of no field near here which can not be drained 2½ feet deep, 2 rods apart, for less than $25.00 per acre. Indeed, most farmers could drain three or four acres each year with almost no cash outlay except for the tile. The draining is at a time of year when little other farmwork can be done advantageously; and I know of few, if any, clay fields which would not be benefited more than that for cultivation and rotation of crops. If even a part of each farm is tile-drained we can follow mixed farming instead of dairying exclusively; and I know whole townships on the Western Reserve which I think are being slowly impoverished by the latter, as at present pursued.

W. I. CHAMBERLAIN.

I may add, that the total average cost per acre of all my drainage, even where less digging and filling than here described was done by team and more by hand, and where prices of labor have been higher than $1.12¼ per day, has been about $23.00 per acre where the laterals are 2 rods apart (33 feet), and about $15.00 to $16.00 where the laterals are 3 rods apart (49½ feet). Tiles are somewhat cheaper now than then, especially the three-inch and four-inch sizes.

PROPER DEPTH FOR TILES.

As already incidentally remarked, I believe 30 inches is, on the whole, the best depth for tenacious and tough clayey soils. I have some 20 acres laid 3 feet, but I now lay only 30

---

* These sizes cost somewhat less now.

inches, and find that seems to do just as well. Below 30 inches our clayey soils, in Ohio at least, are very hard to dig, and *very slow to filter the water*. In the deep black porous soils of Iowa, with deep freezing, 4 feet is none too deep for laterals, and 4½ for mains, and 4 or 5 rods apart will do.

CHAPTER XII.

Conclusion.

The little book is done. It has been lovingly and enthusiastically written. I believe in drainage, especially for the rather flat and slightly rolling but very cold, tenacious, and naturally rather irresponsive and unproductive clayey lands of Northern Ohio and other sections like it. I believe that *tiling* properly, and then *tilling* properly, with manures, fertilizers, clover, and rotation, will make of such soils a very garden for fertility, excellent for wheat, oats, clover, corn, and even potatoes. I should like to persuade my brother-farmers to take the road that has helped me pay my debts and reach a far better net income from the farm. I know that I have written rather as an advocate than as a judge; but I have tried to write with truthfulness and fairness. I have no personal pecuniary interest in leading a single man to lay a single tile, or to plow more, or to use more manures, fertilizers, or clover. I am not financially interested in any way nor in any degree in the manufacture or sale of tiles, clover seed, or tillage implements; but I *am* greatly interested in the great problem of making our clayey farms more productive, and their owners more prosperous. The sandy loams, the limestone soils, the black soils, all reward the labors of the farmer more abundantly at first. It takes more skill to make a good living out of the clayey soils, hitherto given over almost exclusively to grazing and meadows. Tile drainage I believe to be the basis of successful tillage on such soils.

TILE DRAINAGE.

PERSONAL LETTERS OF ADVICE.

Do not ask them. I am a pretty good-natured man, but an extremely busy one. I presume most farmers have little idea how heavy a mail such writers as Mr. Terry, Mr. Gould, Mr. Brown, and myself, have. I find that I can answer personal letters from strangers only with great brevity, and only when a postal card or prepaid envelope, or stamp, at least, is inclosed for reply. If the letters are of general interest I can reply through *The Ohio Farmer*, if so requested. But as to questions of drainage, I have tried in this little book to tell all I know that is necessary for you to know about its principles and practice. If I have failed in the six weeks or so of hard work put upon it, for liberal pay from the publisher, I shall surely fail in the few moments I can afford to donate to you on a postal-card reply. If you have a really difficult problem of drainage, employ a competent engineer, or at least an experienced ditcher, a real expert at the work. I could not advise wisely except by seeing the land.

WHERE TO BUY TILES.

In particular, do not ask me this. I can not tell you, in your locality. Consult the advertisements in your best agricultural and local papers; ask your local freight agent, and write to manufacturers for carload rates delivered at your station, if they are not made near enough to haul with teams. Use your eyes and pen and the mails, or even advertise in your agricultural paper, something like this:

WANTED.--*Carload rates on tile, delivered at ———— Station, on the ———— Railroad.*

Then club with a few neighbors, if you do not want a carload yourself. Begin and tile a few acres near the barn, and then till it "as well as you know how or can learn how;" and as your income increases from that investment, use the surplus funds in draining a littl more each year, thoroughly, and economically, at the times when other farmwork is least pressing.

As promised in the introduction, I have not given here any of what I may call the ancient history of drainage, curious and interesting as is the evolution of any science or art, but of little use now to the average farmer, who wants most to know the best present methods, implements, and materials, and the reasons why they are best. I have also avoided the discussion of curious but really useless side issues and questions. But I have tried hard to touch all points that are of real importance to those who wish to tile-drain portions of their farms, and do it economically and well.

TABLE OF CONTENTS, AND INDEX.

If you wish to learn systematically the scope of the book in a few moments, study the table of contents and the introduction. If you wish to find just where any particular subject is discussed, look for it in the alphabetical index. It has been prepared with care.

TESTING BY TRIAL.

If you wish to learn whether the advice given in this little book is sound, test it carefully and thoroughly on such parts of your farm as seem to need tile drainage.

THE END.

APPENDIX BY A. I. ROOT.

As early as I can remember I had a particular liking for water and every thing pertaining to it. My early home was on one of the hills in Mogadore, Summit Co., O.; and at the base of these hills, beautiful soft-water springs broke forth in abundance; therefore my childhood plays were largely connected with building dams, constructing little water-wheels, carrying water along the sides of the bank in little races, ditches, etc. And during the forty or more years that have passed, running water has had a special fascination for me. Perhaps this may account for the fact of my having had so much to say about springs, artesian wells, irrigating plants, etc., during my travels. The problem of getting water *where* we want it, and *when* we want it, and getting rid of it when we *don't* want it, has been a most interesting one to me. Almost the first useful work I did in my childhood was to make garden; and with my good mother for a teacher, it is not strange that, in later life, I turn again and again to the garden for both relaxation and enjoyment. One of the first problems in gardening was to get rid of the water we did *not* want; and, later on, another problem has been a fond one—the getting of water for irrigation when we *do* want it. This latter, however, will hardly come within the scope of the present work. The former is, however, just what we do want. Just as soon as spring opens we want to get rid of the surplus moisture; and even to this day there are few things I enjoy more than making little open ditches to let the surplus water run off. It is the first work I do in the spring, whenever there happens to be more water on the ground than the drains will take away; for in our locality at Medina, O., the first obstacle to spring work is to get rid of the surplus wetness.

In riding over the country, whenever I see water standing in cat-swamps or sink-holes, doing nobody any good, and damaging the crop, or the chances of a crop, I feel a strong impulse to let the water off. If the owners of such places enjoyed the work as I do, I verily believe they would sit up nights to drain off these eyesores on the land, if they could not manage it otherwise. How I do like to see the water run away, like a liberated bird! and then to witness the dismay of the frogs, turtles, and other denizens of such places, is worth almost as much, if not quite, as to see the wonderful crops which always reward such labor. A swamp, a low wet place, or a springy place, is always a delight to me. If the springy place furnishes water the year round, then it is indeed a little gold-mine—at least, if the water in these springs can be carried somewhere so as to make a nice watering-trough for thirsty horses and cattle, to say nothing of thirsty mankind.

UTILIZING SPRINGY PLACES.

Some time ago I visited a friend of mine who was trying to raise celery. It was suffering from drouth. Not many rods away was some beautiful rich muck; but he could not do anything with it because it was *too wet*. In fact, a stream of water was even then running away from the place. In a very few minutes I had followed the water to a point where it evidently had its source in a clump of rank weeds and bushes; and in a little time we traced it to a place that was even *higher* than his suffering celery. With only a few hours' work his wet place could have been thoroughly drained, and the water from the spring carried to where his growing celery needed it; for the spring was a little higher than any part of the garden. And yet, in spite of all I could say,

I rather think he never utilized the spring, or the ground that it was spoiling by making it too wet. I have met quite a number of cases quite similar.

MAKING CROOKED WATER-COURSES STRAIGHT.

All over our land we find streams of greater or less size running through fields cultivated and uncultivated. The natural course of these streams is zigzag here and there. And that is not all. The greater part of them are constantly *changing* their course. Each freshet starts a new channel. The consequence is, a large part of the field, especially when under cultivation, is wasted. On page 14, Fig. 5, friend Chamberlain graphically describes this state of affairs. See, also, page 62. Such a place was on our own ground when I first settled here ten or twelve years ago. When I spoke of using the low ground for a market-garden, everybody laughed at me. The first thing I did was to cut a straight channel for Champion Brook. We started where it came on my land, and carried it almost in a straight line to a point where it left my land. To do this I was obliged to cut through a bank in one place six or eight feet deep. I soon found I had made a mistake in thinking that I must cut my ditch wide enough for the creek in time of high water. Had I cut a shallow ditch and waited a little, the water would have widened it very quickly, much cheaper than it was done with horses, plow, and scraper. Of course, at every very big freshet I was annoyed by seeing the water break over my embankment and take its old course. It was a good deal like bad habits; with every fierce temptation, resolutions are very apt to be swept away, and the water of passion comes deeper than it did before. Pretty soon it began to be a sort of standing joke, the fight I was having with Champion Brook; and most people thought for a while that I would give way and do as they did—let the brook go where it wanted to. But I didn't. I do not like to fight or quarrel with my *neighbors;* but I do really enjoy a tussle with the elements of *nature*. I enjoy coming out ahead too. Where the water persisted in breaking over into the old channel, I covered the embankment with tin scraps from the tinshop. These were tramped into the clay and gravel. In fact, I first made a horse-path right along the bank, and then a wagon-road; and with the aid of the tin, the dam has become so solid and compact, that, even if the water goes over it in a very high freshet, it does not move it away. At the same time, I kept cleaning out the straight channel a little deeper. It is now wide enough so we can take the big team and a good plow and plow furrows the whole length in the center of the channel up and down, during the first dry time in the spring. Then men come along with shovels and throw the dirt upon the bank, or carry it where it is needed to strengthen the sides of the ditch. But after this, another trouble came in. The water, during each freshet, persisted in cutting the ditch a little wider, and it was encroaching upon my valuable creek-bottom land. To stop this encroachment I bought cedar posts and set them on each side of the creek, in a line six feet apart. The posts were slanted back toward the bank so that the pressure of the earth outside of them should not be too great. Then the tops were secured by a heavy piece of galvanized wire put around each post near the top, the other end being made fast to a stone imbedded in the embankment. This was to hold my fence barricade from being crowded over into the stream, and also to keep the posts in place, should a heavy washout occur during a time of freshet. The posts were cut off in a line where I wished the top of the bank to come, and the pieces cut off were long enough to set in again, so each post made two. Hemlock boards were nailed securely to the posts, and a 2 x 4 cap put on top. Then the dirt was banked back of it. The wagon-road for gathering our crops was close up to this fence or barricade. This work was done last spring. It has endured some very heavy freshets, and yet stands now as sound and solid as when it was put in. Did it pay? Well, you should come down to my creek-bottom ground when we are gathering crops, and see the wonderful growth that we have right in these very low places, where the frogs and toads ten years ago used to hold "high carnival" during the greater part of the season. Underneath the rich black earth of the surface is a porous, gravelly subsoil, such as is usually found, I believe, on such creek bottoms. With the aid of the deep ditch I have described, no underdraining is required on this ground for a good many rods back; and the growth of all kinds of fruits and vegetables here is perfectly wonderful. When I first began to plow up a certain portion of this creek bottom, one of my best men tied his horses to a tree and came back and told me he didn't believe I could ever make any sort of garden in that ground. But I told him to go ahead. But he again declared that I never could get crops from the ground to pay the expense of digging out the stones and roots, and getting it ready for plowing. I bade him go back and stick to his job. The first crop on that ground was onions, and it paid big. After onions we put on the celery, and we have had celery, early or late, on the ground almost ever since. A year or two ago, just for experiment I secured three paying crops on this very piece of ground. First, we picked a tremendous crop of Sharpless strawberries. When the last berry was picked the strawberries were turned under, and a crop of radishes was put on the ground. They grew so quickly that we commenced selling them on the wagon just *thirty days* after the seed was sown. Then we put turnips between the radishes, and had an enormous crop of turnips. And so it has been ever since. This ground is cropped incessantly, from the time the frost leaves it until it freezes up again. This can scarcely be called an open *ditch;* for at some seasons of the year it carries a swift stream perhaps three feet deep and six feet wide.

SURFACE DRAINING.

Friend Chamberlain has had but little to say in regard to

this. In fact, he rather takes the ground that, where the land is properly *under*drained, no arrangements need be made for surface drainage. I have not found this to be the case in market-gardening. As I look out of the window while I write, I see about two acres of ground on a gentle side hill. To prevent washes, I put in lines of tile only 20 feet apart. This *usually* takes away all surplus water about as fast as it falls. With the heavy rains we have here in this part of Ohio, my crops are often injured by being washed out, notwithstanding this drainage. Of course, the tiles run straight up and down hill, or as near it as may be; but during very heavy rains something in the way of surface drainage seems to be still needed. At one time last summer I stood watching the ground from this very window. For an hour or more the tiles seemed to carry all the water, even though it rained in torrents. At last I could see by the looks of the ground that it was fast reaching the point of saturation. Then it began to burst through the furrows here and there, and my nice, fine, heavily manured soil began to rush down into the ditch by the roadside. I had before taken the precaution to have the furrows run *across* the hill instead of up and down, having previously tried the furrows running up and down several seasons. In the middle of the lot is a roadway. This roadway has been sunk down to about the depth the plow goes—that is, the good soil has been removed from the road, and taken to fill up hollows or depressions. Well, a small millrace of water went down this roadway, and that saved the land for two or three rods on each side; but it seemed to me during that heavy rain, that a roadway, say once in six or eight rods, would have been an excellent investment. Then if the ground were plowed so as to make it a little the highest between the roadways, so the furrows would slant from the center each way to the roadway, we should have it. When I visited J. W. Smith, of Green Bay, Wis., I found he had adopted just this plan. His roadways were each right over an underdrain

TILE DRAINAGE.

The land was thrown up into beds between roadways, being a little the highest in the center. These roadways were used for paths for walking, or for running wheelbarrows or running a light wagon to gather crops and distribute manure. Where the roadways crossed, compost-heaps were made, where all useless vines and weeds, trash and rubbish, were stored and composted. This enabled him, in plowing, to clear every thing out of the way of the team. Now, during these excessively heavy rains, the water stood in the *roadways* instead of on the crops. I hardly need tell you that Mr. Smith stands perhaps almost at the head of market-gardeners in the United States. I have visited most of the great gardeners, and I have never seen any thing like Mr. Smith's forty acres.

FILLING UP DEPRESSIONS, ETC.

You will notice, from the above, that I recommend having ground gradually brought into such shape that there will be no depressions where water may stand for even an hour. Friend Terry, in the strawberry-book, urges this quite strongly; and so thoroughly have I become convinced of the importance of it that we had men go over the ground with shovels just before the last plowing, and take the soft earth from the higher portions and throw it into the depressions. This was particularly needed in my creek-bottom land, where the water, by its many years of coursing this way and that, had made little hillocks, and depressions to match, in various places throughout the ground. I noticed that, where water stood for only a few hours on certain crops, it seemed to harm them; and by filling these depressions, and making the general lay of the land so the water would run off in some direction, I avoided having wet places in certain rows. You may object that this is a great deal of trouble; but, please remember that, if the land is once got into proper shape, it is so for ever; and the sight of a field with no depressions where water can stand and make certain spots in the crop yellow and sickly, is worth a good deal to me, and it is worth a great deal to the crop also.

AVOIDING GULLEYS AND WASHOUTS.

A few weeks ago I visited Mr. J. W. Day, the great tomato-grower of Mississippi. He has 400 acres in peach-trees, and is rapidly fitting his grounds for more. All over the South they have terrible times with washouts on the hillsides. This is especially the case where they have the red lands of the Southern States. Since the forests have been cut away, and the grass turned under, many fields have been so cut and gullied they are next to useless. Some of the farmers are making a feeble attempt to fill them up with straw and brush, thinking thus to stop the wash, and divert the water into the proper roadways or channels; but it is a hard thing to stop when once started. About as soon as I looked upon friend Day's gardening and fruit-growing I saw he was running furrows *around* the hills, instead of up and down or crosswise. He had adopted the slope of, I think, one foot in twenty. This carries the water to the nearest open ditch, and prevents it from going straight down hill with a rush. If the descent is much faster than one foot in twenty, it might cut and gulley again; and if more nearly level, it might break across the furrows. Of course, open ditches must be provided, or some equivalent, after the water has been carried about so far. It is somewhat complicated and troublesome, I know, to run these ditches around the hill. But I am sure he makes it pay. His rows of fruit-trees run around the hills in the same way; and in plowing and cultivating, no furrows are made that may suggest to the water (if you will pardon the expression) the idea of cutting across and going straight down hill. This plan is a sort of terracing, only the terraces are run on such a slant as to carry the water. As this land is mostly a gravelly subsoil, no underdraining has yet been done.

Where underdraining is done on low lands, in order to keep

TILE DRAINAGE. 131

them free from silt and mud I have constructed silt-basins. Now, these silt-basins are a bad thing to plow *around*, and they are a bad thing to plow *over;* therefore we cover them with a flat stone, having this stone down so low that the plow will not reach it. The silt-basin is made of a piece of large tile. The underdrain runs in at one side and out at the other. As often as they become filled with mud or silt they must be cleaned out. So far, so good; but if you plow and cultivate right over them, how in the world are you going to find them out? I am ashamed to say that we have some three or four that we can not find, or have not found, for several years. Hereafter I am going to have some landmarks indicating their locality, and have it put down in a book. There is on our land one place where it is underdrained, that has been once taken up and found *completely* filled with mud. This was on our creek-bottom ground.

" WET WEATHER " STREAMS.

What shall we do with streams that exist only in winter time, or during exceedingly wet weather, and disappear as soon as the ground dries off? This is quite a problem here in Medina Co., O. One such low place is on one corner of our grounds. The water goes on to our ground from a sluiceway under a public highway. When I first came on the place I put in it four-inch tile, thinking it would enable me to plow and raise crops over it. As it did not answer I put in six-inch tile. This did tolerably well for three or four years. Finally we had a very wet season, and my raspberries, strawberries, etc., were washed out so repeatedly that I became disgusted. A year ago I bought enough *twelve-inch* tile to go clear through where the stream crosses my land. So far it has carried all the water, and my crops over it are uninjured. As it is some of my very best land, I feel sure that I shall in time get back the money I paid for the twelve-inch tile. In such a place it is quite important that trash be kept from filling up these large tile. At present we made a large silt-basin at the side of the road where the water comes from the culvert, or bridge, and goes into the tile, by placing under the inlet to the tile a hogshead, or very large barrel. All the trash that comes through under the bridge goes down into this hogshead; and when it is full we clean it out. *Floating* stuff is kept out by means of a piece of heavy poultry-netting.

SANITARY TILING.

Much is being said of late about fevers being caused by impure drinking-water. In view of this it behooves every one of us to save the money we pay to doctors (so far as we *can*) and expend it in buying and laying *tiles*. If you can not do any more underdraining, be sure you have plenty of tiles all around your house, your well, and your cistern. Then watch the outlets. Guard them as you would guard your reputation. Last summer, more than $100 was paid out for doctors and medicine because of one spell of malarial fever; and I have decided to invest some more money in tiles in a way that I hope may avert similar attacks in our family. A few days ago I met a poor man who said his doctor-bill was *$280*, and he was in very poor health at the time. He lives in California, where doctors are expensive, especially as he lives several miles from the town where the doctor is. I could not help wondering whether he would not have been *almost* as well if he had not been doctored *quite so much*. Of course, we do not know about these things; but this we do know: Where drain tile will save the necessity of the doctor's visits, it is wiser and better to spend the money in buying drain tile and cutting ditches.

HYDRAULIC ENGINEERS, ETC.

In regard to employing a competent engineer to lay out your ground, as a general thing I believe the man who owns the ground can do his own engineering. Friend Terry's plan is, to pour some water in the ditch and see whether the water will run. If it does not run to suit you, go at it again, and then pour in some more water. Do not lay a tile until the water runs to suit you. I believe friend Terry speaks of drawing water to the field in barrels when it happens there is none ready to test the drain with. While I write I am looking at a spot out of the window where our drains have never worked quite satisfactorily. I told the men who were digging the ditch not to lay any tiles until I got around. Well, the work in the office was quite urgent at the time, and I did not get around as soon as I expected to. They were so sure they were doing it right, they put in the tiles and covered them up. I scolded, and talked about digging them out again. But I let it go; and we have had a wet place there long after the rest of the field is dry. This has happened at every heavy rain since the ditch was made.

This brings us to the matter of poor help. The idea that "any fool can dig" is a big blunder. Friend Chamberlain is exactly right about it. There are only a few people among my acquaintances who can dig well with profit. I remember once upon a time a "tramp" came along and wanted a breakfast. I gave him his breakfast first, and he was to dig until we were satisfied he had paid for it. Well, my impression is that his breakfast would not have been paid for at the present time, had we kept him still digging. He could not do any thing about the ditch without making more trouble than he was worth. Strength is a grand thing in digging ditches; but without brains it amounts to little. In the first place, you want a man who is willing to be taught. For several years I had in my employ an old Englishman. He was a splendid man for most kinds of work; but when laying tiles he would insist on having the bottom wide enough so he could shovel the dirt out with a common barn shovel. Then he would trim down the sides until they were smooth and true. His ditch was very handsome, and would have been just the thing if he were going to lay tiles a foot square. But it is a great blunder to dig such a big trench for round tiles only $2\frac{1}{2}$ or 3 inches in diameter. I had to give him up.

He would *not* dig as I wanted him to, but insisted that it was cheaper to have a ditch so wide that a man could walk back and forth, and turn around with ease at the bottom. I found a stout German who was so new from the old country that he could hardly speak a word of English. But he would do exactly as I said ; and when I showed him how, his every effort seemed to be to do just as *I* wished to have him do, instead of following his *own* notions. His ditches worked all right, and he soon learned to make very good speed. I would exhort our readers to read again and again the instructions on page 86, and all along there*. Don't give up and go back to your old-fashioned ways. Keep on until you can do it all exactly as friend Chamberlain directs, and you will very soon have good reason to be glad you did so.

The one who succeeds in underdraining must be more or less of an enthusiast in the work ; and this seems to be true of all industries, especially those of a rural nature. I was pleased to hear J. M. Smith say, that, after his drains were all made, he watched anxiously for a heavy shower, to see if they would work as he had planned to have them do ; and so anxious was he that he did not wait for the rain to be over, but with gum boots, rubber coat, and umbrella, he went out through the rain to one of the silt-basins, where a large number of drains were centered, and he was delighted

---

* Since the above was in type *I* sent a big strong man who has done ditching, and thought he knew how, out into the lot to dig where I had previously drawn a string for him to go by. It was two or three hours before I got ready to come around and inspect. I found him making but little progress in trying to dig the hard clay with a *pick*. I took the spade given him at first, and taught him to throw out the dirt a full spade depth almost as fast as he could handle his spade, and this, too, in ground so hard he thought he would have to use the pick. Had he tried to push his tile-spade into the ground with both edges in the clay it would have been impossible, even if he had put his full weight (200 lbs.) on top of the spade; but by putting only one edge in the clay, leaving one edge clear, he managed it easily, and the dirt dropped from his spade without a bit of trouble; whereas, by the old plan the clay would stick to the spade so as to have to be cleared off with a trowel almost every time. Why, since I have learned how to do it as friend Chamberlain does, I find it just fun to dig in the hardest ground, and the labor isn't severe and exhausting either.

to see every one of them pouring its proper proportion of water into the basin, while the main outlet leading from this, carried the water away. Each line of tile seemed to be fully adequate to take the water, even from the heaviest shower. After you have worked and planned, perhaps many months, on a piece of engineering like this, what a satisfaction it is to see it a success in every particular! Friend Smith had a problem on his hands of more than usual difficulty, for his field of forty acres is very nearly level, and it is also but a very little higher than the water in the lake which is all around him; therefore his drains must be laid very accurately, on an even, true grade; for, as there was but little fall to be secured, there could be no very great variation in the way of ups and downs.

DEPTH OF TILE, ETC.

I quite agree with what friend Chamberlain says in regard to the depth that tile should be laid. In our locality most of the upland is a stiff retentive clay. If you dig a hole in the ground anywhere, it will fill with rain water, and the water will stand there a long time. On account of this, I think I have several times placed our tiles too deep to give the best results. It has been suggested that, after a year or two, the water would find its way down to them. This may be; but, all things considered, I think I should prefer them not lower than $2\frac{1}{2}$ feet from the surface, as a rule. In order to get an even grade, I would, of course, go down in places, 3 or even $3\frac{1}{4}$ feet; and in other places we might lay them as near as 2 feet from the surface. In my recent travels through Washington, Oregon, California, and Arizona, in going among the farmers and fruit-raisers I have been greatly astonished to find how little the teachings of one locality would answer for another; and not only is this true, but circumstances are so different, sometimes, that, in traveling only thirty or forty miles, their methods must be greatly modified and changed. In the great West, underdraining is, as a rule, almost unknown. The desert soils where irrigation is practiced, are so loose and porous that the ground is really underdrained by nature. In fact, in many places they have great trouble in conveying the water in irrigating-ditches, the soil being so porous that the greater part of the water is lost before it goes to where it is wanted. In fact, at Riverside they are cementing the bottom and sides of the irrigating-canals, and this is being done very largely. Now, in such soils, and especially where there has never been a surplus of rains, except for a brief period in the winter time, underdraining would seldom pay expenses. And this leads us to a method of getting an outlet for underdrains that has already been touched upon. In the vicinity of Mammoth Cave, water never stands in hollows or depressions. In fact, in passing along on the railroad we see hollows between the hills, without number, and no visible outlet anywhere. After very heavy showers the water will sometimes stand in these basins for a few hours; but it usually goes down through the porous soil and through the porous rock almost as rapidly as it collects in the basin between the hills. In some cases I saw considerable streams go down into such a hollow in the hills, and disappear in a sink-hole that resembled a well without any bottom to it. The water had probably found a passageway into some of the caverns that underlie that large tract of that Mammoth Cave country. Now, in localities where there are no caves, there are frequently porous strata of gravel that act in much the same way; and in such places an outlet for underdrains may be found by simply digging a well down into this gravel. Even here in our hard clay subsoil, with soapstone underlying, we drilled a well that we found, by actual test, would take quite a stream of water, when it was once turned into it, immediately out of sight. A good many times it may be worth while to experiment in this way, where it is desirable to get an outlet for surplus water. On page 104 we are told how friend Terry ran the water into a gravel-bank, and thus saved the expense of making a deep cut.

OUTLETS THAT ARE LIABLE TO BE FLOODED BY OVERFLOWS, ETC.

There is one spot on our creek-bottom grounds where we get an excellent outlet for standing water by running a tile into the deep cut before spoken of. Now, this works all right while the water is low in Champion Brook; but when it rises nearly to the top of its banks during a freshet, the same tile that had been taking the water away lets it back on to the low land, and makes a pond. To remedy this we have at the outlet a check-valve made of wood, with leather hinges. Ordinarily the valve stands a little open to let the water out. When the water rises, however, it strikes this hinged door, or valve, and closes it. Of course, this stops the water for a time from getting out of the underdrain; but when the creek goes down it lets the water out as usual. We should prefer, of course, an outlet that would never be covered and subject to back-water; but under the circumstances the valve is better than to have the tile permanently open. A valve to open and close by hand has been suggested; but as this requires the owner to go out in the rain to manipulate his valves, and would often fail to receive the attention it should have, we consider the automatic valve preferable.

DANGER FROM STANDING WATER.

Perhaps sufficient has been already said in regard to the damage a little surplus water may do if allowed to stand for only a few days But I wish to mention a circumstance of my boyhood. We had a very thrifty young cherry-tree, one of the very finest I have ever seen, standing not very far from the house. Before we had our underdrains properly fixed, a wooden spout carried the surplus from the kitchen out to an open ditch. During a wet time in the spring this wooden spout became disarranged so the water for two or three days poured down near the cherry-tree mentioned. In fact, it had made a little puddle around the tree before anybody had noticed it. For fear it might receive injury, an open ditch was made out to the main ditch, to take the surplus water away. But we were too late. The tree, although it had commenced to put out its buds with wonderful thrift and luxuriance, came to a standstill and died. It seems to me the water could not have been around the roots more than three days. I have seen evergreens on our present ground killed in the same way. They were planted through a depression, where water stood occasionally during very heavy rains. A succession of heavy rains kept this place full of water for perhaps 48 hours. Then a ditch was dug, and some tiles put in; but it was too late—they were dead—died from drowning, i. e., being held under water longer than they could stand it. This brings me to one of the main points in regard to surplus drains. On our rich market-garden ground, where we have put on manure year after year at the rate of forty or fifty loads per acre, the ground has not only become very rich, but very soft, and easily worked. Well, in gathering the crops we are sometimes obliged to go over the ground when it is pretty wet. It gets tramped down hard, to the great detriment of the growing crops. Our plant-beds, however, are never tramped on. As they are only six feet wide, one can reach from the edges to the center of the bed, and the boys are forbidden to ever set foot on this thoroughly tilled and heavily fertilized soil. The consequence is, we get crop after crop from these beds, with very little digging up more than is required by the rake. The beds being only six feet wide, and paths between them, they make the most perfect surplus drainage. In fact, the beds are raised perhaps four inches higher than the path. This, of course, makes them dry out quicker during a very dry time. But here comes in our big windmill, with its 1600-barrel tank. Hydrants are located among the plant-beds about 100 feet apart, so 50 feet of hose, with the sprinkler attached, enables us to see that the crops on our beds never suffer from a lack of water; neither is the soil ever tramped down so as to prevent the roots from being able to "breathe" through the loose soil. It may be that this is not exactly the way to express it, but you all know what I mean. I think one reason why we can not compete in fruit-growing with California is because we have not the loose porous soil they have. We can, however, have it more loose and porous than it now is by keeping it at all seasons of the year free from standing surplus water. In our locality I have often seen a field plowed, harrowed, rolled, and cultivated until it was light, loose, and fine, and in just the trim for putting in the seeds. Very likely it had been just rolled the last time preparatory to running over the seed-drill. Now, if, at this stage of proceedings, a tremendous rain sets in, so that the whole field is made like mud, it will settle down so hard and compact that the chances for a good crop are just about ruined. The only real remedy for such a state of affairs is to wait until the ground is dry enough to plow and harrow and roll again.* If, however, the field had been thoroughly *underdrained*, the rain does comparatively little damage. With surface drains, such as I have mentioned, the damage is still less; but the owner must keep off the ground with his team and tools until it is sufficiently dry to work.

TOOLS FOR UNDERDRAINING.

Our friends may have noticed that some of the tools on page 82 have not been described or mentioned. This is because some of them are of my own selection. Fig. 2 is a Dutch hoe. One of my men, fresh from the old country, brought such a one along with him. The others laughed at it, and said Fig. 8 was a good deal better, and answered every purpose. When it came to filling up the drains, however, our German friend got his hoe (Fig. 2) and soon convinced

---

*The delay, especially if it *keep on* raining, may be about as disastrous as to go ahead.

them that, in his hands at least, it was a wonderful tool, and therefore the boys took a liking to it for many other purposes. It will dig up ground many times in a way to make it quite a substitute for spading, and do it easier and quicker. Fig. 5 is a cleaning-out spade, and is very handy in many emergencies. Fig. 3 is used as a substitute for Fig. 1. It is mostly used for cleaning out the bottom, and grading it for laying small tile, and I like it rather better than Fig. 1. The adjustable joint in Fig. 1 is a good thing if it were not so apt to be loose and rattling. It always makes me nervous when any tool is the least bit loose on the handle.

On page 91 we have illustrated and described a three-tined ditching-spade. I am ashamed, however, to think that our artist did not succeed in making a better-looking cut of it—like Fig. 6, on page 82, for instance. Well, after what friend Chamberlain had written on page 92, I purchased one of these spades and sent it to him to try. Here is his reply in regard to it:

SKELETON SPADE.

Mr. A. I. Root—Dear Sir:—I express you the skeleton spade to-day. On the whole the regular ditching-spade seems to me better than the skeleton for my land. This particular one has too narrow a blade, except for bottom course, and the shoulder is so narrow that the foot slips; also, the ribs, or tines, bend too easily. They were bent when it came, and I straightened them with care, but they are not stiff enough, even for a skilled ditcher, in our soil. I would not, however, decide wholly against them until I try the wider and shorter size in a very wet and sticky time and place. Thus far my men and I like the solid or regular spades best; but in Iowa soils the skeleton spade is just the thing.
Hudson, O., March 15, 1892.   W. I. CHAMBERLAIN.

We have used the spade a little, but not very much, however. But I am inclined to think that, where it is properly made, as friend Chamberlain suggests, i. e., made for our soil, it will clear better and cut easier than the one figured on page 82, No. 6. I should be glad to hear from friends who have used the three-tined spade. It seems to me the idea is a progressive one.

CONCLUSION.

Before closing I wish to say that I have visited friend Chamberlain at different times during his work. He is not very far from friend Terry, so when I go to see one I take in the other also. I have seen that beautiful orchard, and noticed the great difference in the trees that are on the tile-drained land, and I also saw that Baldwin apple-tree, page 49, just before they were going to gather the fruit. It was a sight, I'tell you, and I never in my life before saw such a lot of beautiful, round, smooth, perfect apples as I saw at friend Chamberlain's. One of the most interesting sights to me was the item mentioned and illustrated on pp. 44, 45, 46; namely, the idea that *tiles* helped the *crops* but *hindered* the weeds. When I was first told it, I felt like making light of friend Chamberlain's enthusiasm, which I thought then was running away with him. He took us first where we saw a heavy growth of clover and timothy, with no weeds or plantain. Then we looked at the ground that had not yet been tiled, and I soon took in the state of affairs. There are certain kinds of weeds that we expect to see appear on poor, hard, dry ground. In fact, they are seldom seen elsewhere. There is but little grass or clover or any thing else on such ground, and I had always supposed that the rank, hardy weeds had killed out the grass. If I am correct, however, the real state of affairs is this: The excessive *wetness* first kills out the timothy and clover in the same way it killed that cherry-tree and our evergreens. When the timothy and clover are out of the way, the hardy weeds that can *stand* the wet come in and cover the ground; therefore we have a rational and reasonable explanation given for the fact that we *get rid of weeds* by thoroughly underdraining our land.

This book, like the various books written by friend Terry, is not theory or *book* farming, like some books that have been written on agriculture, for it is only a description of what I have seen repeatedly, and what you too, dear-reader, may see for yourself if you feel like taking the time to visit Terry and Chamberlain. Although they are busy men, and their time is exceedingly valuable, I think I am right in saying they are ready to sacrifice almost any thing, where they can be of use in making themselves really helpful to the average farmer. In fact, they both have been giving their lives to this work, and I suppose they expect to give them to the end. I do not mean by this to say they give their time right along for nothing, and without pay. If we American people allowed them to do so we should be ashamed of ourselves.

If this little book has been the means of giving you more faith in and love for farming; more faith in and love for your fellow-men, and more faith in and love for the great Creator, who gave us our land with all its grand and glorious possibilities, then shall we be satisfied, and feel that our labor has not been in vain.

Your friend, A. I. ROOT.

Medina, O., March 17, 1892.

TABLE OF CONTENTS.

Chapter I.—Introductory.

The scope of the book.—A primer in size and conciseness.—On *tile* drainage, not all drainage nor all related subjects.—Drainage a progressive art.—Much of the past now discarded. Tiles have superseded all other materials for underdrainage.—Round tiles best.—Author's practical experience and observation.—Tiling not so difficult as sometimes represented, yet needs an engineer for difficult problems. Pages 3 — 6.

Chapter II.—Why do we Tile-drain Land? the Theory.

Do all lands need tiling?—Horace Greeley's dictum.—Underdrainage better than surface drainage.—What is "surplus water"? Upon and in the soil.—Roots need air.—Capillary attraction in lamp-wicks, sponges, soils.—Air-spaces in soils.—Capillary water and hydrostatic water.—Porosity and slight filtration of "loose" soils.—Typhoid germs.—Artesian wells.—Capillarity and hydrostatic pressure oppose each other.—Scientific reasons for removing surplus moisture down through the soil: 1. Fits fields for less crooked tillage; 2. Removes surplus in as well as on the soil; 3. Saves loss of fertility; 4. Even adds fertility; 5. Helps warm the soil; 6. Lengthens the season of tillage and growth of crops; 7. Increases extent of "root pasturage;" 8. Helps disintegrate and "fine" the soil; 9. Diminishes "winter-killing" of wheat, clover, etc., by "hoar frost" or "stool ice;" 10. Diminishes effects of drouth; 11. Often diminishes flood damage; 12. Improves the healthfulness of drained regions. Pages...7 — 28

Chapter III.—Why do we Tile-drain Land? The Facts. Does it Pay?

Author's first cobble-stone drain.—First tile drain.—First thorough tile drainage less than $23 per acre, and first wheat crop $46 per acre.—Naturally gave the "tile-fever."—Actual effects of tiling: 1. Upon orchards; "died of wet feet;" 2. Upon wheat; 3. Upon clover; 4. Upon value of manures and fertilizers; 5. Upon *permanence* of fertilizers and crops, and upon weeds; 6. Upon barn room; 7. Upon the fruitage of apple-trees. Pages..28 — 50

Chapter IV.—Does Tillage Pay better than Grazing?

Not exclusive tillage.—Combined with stock-keeping.—Agricultural development of the human race.—The savage state.—The nomadic.—The agri*cultural*.—The horti*cultural*.—Population of Ohio per square mile.—Of Belgium.—Exclusive grazing in Ohio.—Tillage with stock-raising.—The tiling of clayey soils a necessary *basis* for tillage and rotation.—Manures, fertilizers, clover, rotation, and good tillage must follow to win success.—

Brief history of drainage on author's farm.—Effects of tiling, tillage, tree-planting, etc., upon the rainfall.—Soils that most need drainage, thorough and partial. Pages............51 — 62

CHAPTER V.—WHERE TO DRAIN.

General localities and kinds of soil.—Cat-swamps, swales, etc.—Location of mains and laterals in a system of thorough drainage.—Mains follow but somewhat straighten the dry brooks and swales.—The direction of laterals. Two rules.—Criticisms of.—Discussion of.—Illustrations, Figs. 18, 19.—How the surface-water seeks the drains.—Dead-furrows?—Rapidity of absorption and filtration. Pages...............................62 — 71

CHAPTER VI.—WHEN TO DRAIN.

1. When can we afford it? Debt for it? Sell part to tile the rest? Gradually, year by year, and economically.—2. Best seasons of the year. Tiling in winter, late fall, and early spring.—Advantages of.—Preparation for.—Details of in next chapter. Pages..... ....71 — 74

CHAPTER VII.—HOW TO DRAIN; THE TILES.

The tiles.—Shape.—Material, and hardness of burning.—Porosity.—Experiments concerning.—The water enters drains how and where?—Hard tiles more surely durable.—"T joints" and "Y joints" needed for hard tiles.—Better for any kind of tiles. Pages...74 — 78

CHAPTER VIII.—HOW TO DRAIN; THE TOOLS.

Hand tools or machines?—Field trials of horse and steam power machines. Greatly hindered by stones in drift soils or bowlder clay.—Not suited to winter work.—But winter the time of leisure for drainage.—Shall the farmer buy a digging-machine as he does a mower or twine-binder?—Shall he hire one?—Hand tools; spades; scoops, or crumbers; groove-cutters; span-level; filling hook and fork; crowbar, pick, shovel, etc. Pages.78 — 84

CHAPTER IX.—HOW TO DRAIN; THE MANIPULATIONS.

Locating the drains.—Wet-weather observations.—Drainage engineer; grade-stakes, map, etc.—Winter drainage.—Beginning to dig.—Where? How?—*The main* first laid.—"T" and "Y" joints, when laid. The laterals.—The digging.—Skill by practice and thought.—"Burying" the spade.—Keeping one side-edge in sight.—How to sink the spade; and preserve the true grade; and use the "crumber."—The foot-iron.—Placing the earth so as to fill in easily.—The curved motion in lifting the earth.—Centrifugal force helps keep the earth on the spade.—Skill or knack required to dig a clean, true ditch rapidly, easily, and well.—The three-tined ditching-spade for stoneless, mucky, sticky soils.—Establishing close grades.—"Sighting-stakes" and "boning rods," Fig. 25.—Cutting the groove for the tiles, Fig. 26.—Depressions in the drains endanger stoppage.—Use of the span-level in getting grade.—Of soil-water.—Of water brought on purpose.—The work to endure for centuries, if well done; for about two years if not well done.—Laying the tiles.—Covering them.—Tramping the earth.—How the water enters.—The upper ends of laterals.—Protecting outlets from land and water vermin, reptiles, etc., by screens.—Have as few outlets as possible. Pages....................84 — 100

CHAPTER X.—HOW TO DRAIN; SPECIAL PROBLEMS.

Map of thorough drainage of 36-acre field on author's farm, Hudson, O.—Handling flood-water from land further up the slope.—Ponds as balance-wheels.—"Catch-water pond" and "catch-basin."—"Contour lines."—Study of the map.—Drainage into reservoirs sunk into gravel subsoil.—By mains running into gravel-knolls, and with no outlets.—Protecting joints from entrance of quicksand or "silt."—Establishing grade by line instead of boning-rods and sight-stakes.—Quicksand-pockets; boards to prevent the tiles from sinking; silt-basins, or manholes.—Boards under tiles at outlets.—Tiles rather than stones, even where stones are plentiful and of good shape.—But if you use stones, lay a clear channel for water.—Roots in tile drains.—Not unless the tiles are wet when the soil is dry.—Former belief as to porosity of tiles changed by later careful experiment.—Roots of aquatic trees.—Conveying spring or swamp water through dry soil.—Slop-drains, water-closet and sink-drains, etc.—Drainage of low lands by steam or wind engine pumping.—Letter quoted.—Scarcely within the scope of this book. Pages ....100 — 112

CHAPTER XI.—ESTIMATES; SIZES OF TILES; COST OF DRAINING.

Size of tiles to drain given areas. Nature of the problem.—Sudden and heavy rainfalls.—English estimates too small for the United States.—The "factor of safety."—Actual rainfalls on author's farm.—Sizes of tile that "handled" it without surface-wash.—Author's rule for size. Rule by C. G. Elliott, in his "Practical Farm Drainage."—Rule by Geo. E. Warring, Jr., in his "Draining for Profit and Draining for Health."—Tables by French and by Klippart in their books on drainage.—Present tendency toward large tiles wise.--Author's conclusions.—Cost of drainage if done economically.—Actual cost itemized for 15 acres.—Proper depth in tenacious clays in moderate climate.—In more porous soils and colder climates...Pages......112 — 120

CHAPTER XII. CONCLUSION.

Good tillage should follow tile drainage.—The author an advocate of drainage, but not from personal, pecuniary interests.—Has tried to give theories correctly and facts truthfully. Personal letters not practicable.—Where and how to buy tiles.—Begin the work of tiling gradually and economically (if your land needs it), and be guided by observed results. How to find what is in this book, or what it has on any particular point. Test the book's advice by trial. Pages........................120 — 122

INDEX.

Air-spaces in soil 9, 10, 22
Attraction, capillary 8, 10
Artesian wells 12
Agriculture, development of among men 51, 52, 53
    " supports a greater population than grazing 53, 54, 55
Belgium, population to the square mile 53
Boning-rods, or sighting-stakes 92, 93
Catch-basin 102
Country Gentleman, quotations from ... 6, 96, 108, 112, 117, 118
Capillary attraction in soil, sponges, etc. 8, 9, 22, 23
    " water 10
Cook, Prof. A. J. 1
Camp, H. B. 110
Clover saved from winter-killing by tile drainage 22
Drainage—tiles superseding other material 1
    " a progressive science 1
    " down through the soil, why better than off from it 13, 14, 15
    "   "   " aids in pulverizing it 21
    "   "   " prevents winter-killing 22, 23
    " diminishes the effects of drouth 25
    " (sometimes) diminishes the suddenness and violence of floods 26
    " improves the health of a region 27, 135
    " author's first experiment in cobble drainage 29
    "   "   " tile   " 29, 30
    " effects of on orchard trees .... 31—38
    "   " wheat 38, 39
    "   " clover .... 39—41
    "   " value of manures and fertilizers 43
    "   " permanence of timothy seeding and on weeds, 46
    "   " barn room 47, 48
    " tools for (see Tools) .... 78—81
    " of bogs 103, 104
    " by steam pumping instead of gravity outlet 111, 112
    " cost of—estimates .... 117—119
Drains (tile), location and direction of mains 63
    "   " laterals 64
    " straight down the slope or not .... 64—70
    " dead-furrows in connection therewith? 68
    " rapidity of absorption and filtration by soil into 70, 71
    " when to lay them (the question of affording them) ... 71, 72
    "   "   " (best season of the year) 72, 73, 74
    " construction of—the tiles, material, shape, hardness, etc 74
    " porosity of tiles; transmission of water .... 74—76
    " weight of tiles 77
    " beginning the digging (where?) method of digging 85
    " establishing close grades 92, 93, 106
    " cutting the groove and laying the tiles 94
    " protecting the outlets 96
    " depth of 108, 117, 119, 135
    " distance apart 117, 120
    " without open outlets 104
    " roots in 108, 109

TILE DRAINAGE.

```
Debt with relation to drainage .................................5, 120
Dry-earth mulch................................................23, 24
Drouth diminished by drainage........................................25
Engineer, employment of in drainage.........................5, 92, 134
Evaporation, a slow and cooling process.........................17, 18
Emerson, R. W., quotations from..................................1, 21
Elliott, C. G......................................................115
Fertility preserved and increased by tile drainage..............15, 16
Frost action, effect on soils and crops.........................22, 23
French on drainage, quoted..........................21, 98, 99, 116
Greeley, Horace, his dictum about drainage...........................7
Growing season lengthened by drainage...............................19
History of drainage—why not given here..............................4
Hydrostatic water...........................................10, 11, 23
Hoar frost explained............................................22, 23
Heat makes the air hold more moisture...............................24
Health improved by drainage..........................27, 28, 132
Hall, W. B........................................................108
Johnston, John, reference to.........................................5
Johnson, B. F., views on drainage...................................30
Joints of tiles covered to exclude silt............................105
Killmer, Charles S.................................................111
Klippart on drainage..............................................116
Leveling for drains........................84, 85, 92, 93, 94, 106
Laterals, upper ends of............................................96
   "      direction of............................................64
   "      ending in open ditches.................................107
Main drains, location of...........................................63
Motley, J. L., reference to.........................................4
Map of thorough drainage..........................................101
Mulch of dry earth retards evaporation..........................23, 24
Main drains, direction of..........................................63
   "       "   ending in gravel beds with no outlet.............104
National Stockman referred to................................6, 103
Ohio—The Ohio Farmer, references to...................6, 105, 121
   "   meteorological bureau......................................15
   "   population to the square mile..............................53
   "   exclusive dairying in...................................53, 54
   "   mixed or diversified farming in.........................54, 55
Outlets, protecting from vermin....................................96
   "    ending in gravel subsoil...............................104, 134
Primer, this book designed to be in size and conciseness............3
Pulverization helped by tile drainage...........................21, 22
Paper to cover tile-joints........................................105
Personal letters of advice........................................121
Population to the square mile as affected by tiling and tillage..55, 57
Quicksand.........................................................106
Rainfall on author's farm.......................................15, 16
   "     increased by drainage, tillage, and tree-planting....57, 58
Rural New-Yorker, reference to......................................6
Roots of plants need air........................................9, 10
   "      "    grow deeper in tiled soil......................19, 20
   "    obstructing drains...................................108—111
```

Saturation, drainage lowers the line of 22, 23
Swan, Robert J., reference to 5
Sisson Brothers, reference to 5
Stockman, National, reference to 6, 103
Surplus water, how to remove and why .... 7, 11, 15, 16, 17, 18—31, 137
Soils, what ones need tiling 7, 59, 62, 136
" temperature of affected by tiling 16, 17
Sighting-stakes, or boning-rods 92, 93
Superphosphates more effective on tiled land 55
Scope of the book .... 1—5
Silt or sand working into tile drains 105
Silt-basins 106, 107
Storrs & Harrison Co. referred to 106
Stones, use of for drains 5, 29, 107, 108
Sizes of tiles, rules for determining size needed, and given ...... 112—116

Thawing, a cooling process 17
Tile drainage—see Drainage.
Tiles superseding other material for drains 3, 4
  " round, superseding other shapes 4
  " where to buy 121
  " durability of and material for .... 4, 74—77
  " proper sizes for various areas .... 112—116
Tiling and tilling not to be confounded 6
  " what soils need it 7
Tillage of surface to retain moisture .... 23—25
  " does it pay better than grazing? 51
Terry, T. B., references to 1, 5, 21, 60, 104
Typhoid germs saturating soil and polluting spring water 11, 12
Trowbridge, W 105
Tools for drainage—hand or machine? 78
  Trials of machines 78, 79
  Stones interfere with work 78, 79
  Mud and frost interfere with .... 79—81
  Common plow for deep furrow the only team-machine used 79
The hand-tools, spade for first course 81
  Bottoming-spade 81, 139, 140
  Scoop for first course 81
  The bottoming-scoop, or groove-cutter 82
  The span-level 83
  The tile-hook 83
  The four-tined filling-hook 82, 83
  The three-tined ditching-spade 91
  The foot-iron 89
  Shovel, crowbar, pick, iron rake, etc 84
  Sighting-stakes and boning-rods 93
Townshend, Prof. N. S 110

Vermin in drains .... 96—99
Velocity of flow in tile drains affected by grade and friction 113

Wet feet, plants and trees dying of 8
Winter-killing of wheat, clover, etc., diminished by drainage 22
  " " explained 22, 23
Wind, why more drying than a still air 24
Western Reserve, much of the soil benefited by drainage 61, 62
  " " author's experience with its soils 62
Waring, G. E 106, 115

A B C OF STRAWBERRY CULTURE.

FOR FARMERS, VILLAGE PEOPLE, AND SMALL GROWERS.

A BOOK FOR BEGINNERS.

BY T. B. TERRY.

The above book, by Terry, with some additional remarks by A. I. Root, is, at the present time, creating an enthusiasm and interest in strawberry culture never known before. It is a book of 144 pages and 52 engravings, and it is fully up to the times.

Price 35 Cts. By Mail, 40 Cts.

Published by A. I. Root, Medina, Ohio.

GLEANINGS IN BEE CULTURE,

A 36-Page Semi-Monthly.

$1.00 Per Year.

Each issue is printed on the finest of book paper with the best book ink, handsomely illustrated with a large variety of original engravings. Sample copy free on application.

—THE MANUFACTURE OF—

BEE-KEEPERS' SUPPLIES

IS OUR SPECIALTY.

A 52-page catalogue of every thing pertaining to bee culture will be sent for your name on a postal.

A. I. ROOT, MEDINA, OHIO.

THE A B C OF CARP CULTURE.

Second Edition, Revised and Entirely Re-written.

A COMPLETE TREATISE ON

THE GERMAN FOOD CARP.

Including Plans and Specifications, and Fullest Instructions for the Construction of Ponds, and Every Thing Pertaining to the Business of Raising Carp for Food.

Illustrated by Many Fine Engravings

By A. I. Root and George Finley.

PRICE 35 CENTS; BY MAIL, 40 CENTS.

THE A B C OF POTATO CULTURE.

By T. B. Terry, of Hudson, Ohio.

How to Grow Them in the Largest Quantity, and of the Finest Quality, with the Least Expenditure of Time and Labor.

Carefully Considering all the Latest Improvements in this Branch of Agriculture up to the Present Date.

FULLY ILLUSTRATED.

TABLE OF CONTENTS.

Soils and their Preparation.—Manures and their Application.—When and How Far Apart Shall we Plant?—Shall we Plant Deep or Shallow? Shall we Plant in Hills or Drills?—How to Make the Hills and Fill them. —Selection and Care of Seed.—Cutting Seed to One Eye.—Planting Potatoes by Machinery.—Harrowing after Planting.—Cultivating and Hoeing.—Handling the Bugs.—The Use of Bushel Boxes.—A Top Box for the Wagon.—Digging.—Storing.—What Varieties Shall we Raise?— Potato-Growing as a Specialty.—Best Rotation where Potatoes are made a Special Crop.—Cost of Production, and Profits.

PRICE 35 CENTS; BY MAIL, 38 CENTS.

PUBLISHED BY A. I. ROOT, MEDINA, OHIO.

THE WINTER CARE OF
HORSES AND CATTLE.
THE MOST HUMANE AND
Profitable Treatment.
BY T. B. TERRY.

Although the book is mainly in regard to the winter care of horses and cattle, it touches on almost every thing connected with successful farming.

SHELTER, COMFORT, FEEDING, EXERCISE, KINDNESS, DIFFERENT SORTS OF SEED, AND A FULL TREATISE ON THE MOST ECONOMICAL WAY OF SAVING MANURE.

A full description of Terry's model barn is also given.

PRICE 35 CENTS; BY MAIL, 40 CTS.
PUBLISHED BY A. I. ROOT, MEDINA, O.

MAPLE SUGAR AND THE SUGAR BUSH.
—BY—
PROF. A. J. COOK.
—AUTHOR OF THE—
Bee-keepers' Guide, Injurious Insects of Michigan, etc.

The name of the author is enough of itself to recommend any book to almost any people; but this one on Maple Sugar is written in Prof. Cook's happiest style. It is

✸ PROFUSELY ✸ ILLUSTRATED, ✸

And all the difficult points in regard to making the very best quality of Maple Syrup and Maple Sugar are very fully explained. All recent inventions in apparatus, and methods of making this delicious product of the farm, are fully explained.

PRICE 35 CENTS; BY MAIL, 40 CENTS.
PUBLISHED BY A. I. ROOT, MEDINA, OHIO.

THE A B C OF BEE CULTURE.

BY A. I. ROOT.

A Cyclopedia of Every Thing Pertaining to the Care of the Honey-Bee.

This is a cyclopedia of 400 pages, and is beautifully illustrated by over 300 engravings, many of them full page. Some of the latter embrace a view of the apiaries of some of our largest and most successful bee-men. The whole work is elegantly bound in cloth, 7 inches wide by 10½ inches long, and embossed on side and back in gold. It would be an ornament on the center-table of any bee-keeper's home.

Since the first issue in 1877 its average sale has been over 200 copies per month, and the sale has been steadily increasing from the first. Prices: Neatly and strongly bound in cloth, by mail, $1.25; by epresss or freight with other goods, $1.10.

Merrybanks and His Neighbor.

By A. I. Root. This is the title of a little book of 210 pages and 68 illustrations. It narrates the alternate failure and success of a beginner who ultimately, through much tribulation, becomes a successful bee-man and a power for good in Onionville. Appropriate original cuts, many of them humorous, are interspersed here and there, representing some of the droll experiences which a beginner with bees sometimes passes through. Besides bees, it talks of other rural pursuits, such as gardening, maple-sugar making, etc. It has a good deal to say about our homes, and more particularly one home which was started upon a sandy foundation, but eventually became builded upon the rock Jesus Christ. The book is full of instruction; price 25 cents; 3 cents extra when sent by mail.

What to Do, & How to be Happy While Doing It.

The above book, by A. I. Root, is a compilation of papers published in GLEANINGS IN BEE CULTURE in 1886, '7, and '8. It is intended to solve the problem of finding occupation for those scattered over our land, out of employment. The suggestions are principally about finding employment around your own homes. The book is mainly upon market-gardening, fruit culture, poultry-raising, etc. I think the book will be well worth the price, not only to those out of employment, but to any one who loves home and rural industries. Price in paper covers, 50 cts.; cloth, 75 cts. If wanted by mail, add 8 and 10 cts. respectively.

A. I. ROOT, MEDINA, O.

NOW IN PRESS.

—A TREATISE ON—

TOMATO CULTURE,

—BY—

J. W. DAY, OF CRYSTAL SPRINGS, MISS.

WITH AN APPENDIX BY A. I. ROOT, ADAPTING IT TO TOMATO CULTURE IN THE NORTH AS WELL AS IN THE SOUTH.

This little book, which we expect to have about the size of the present manual, is interesting because it is one of the first rural books to come from our friends in the South. It tells of a great industry that has been steadily growing for some years past; namely, tomato-growing in the South, to supply the Northern markets. The little book, which is to be very fully illustrated, gives us some pleasant glimpses of the possibilities and probabilities of the future of Southern agriculture. Even though you do not grow tomatoes to any considerable extent, you will find the book brimful of suggestions of short cuts in agriculture and horticulture, and especially in the line of market-gardening. The price will probably be 35 cents; by mail, 40 cents.

A. I. ROOT, - - MEDINA, O.

THE DRAINAGE JOURNAL,

A 48-PAGE MONTHLY,

—DEVOTED TO—

FARM DRAINAGE, WHEN AND HOW TO DRAIN, ROAD IMPROVEMENT, AND MANUFACTURE OF DRAIN TILE.

NOW IN THE THIRTEENTH YEAR OF ITS PUBLICATION.

PRICE ONE DOLLAR A YEAR.

SAMPLE-COPY TEN CENTS.

Address DRAINAGE JOURNAL, Indianapolis, Ind.

The beautiful cuts on the back and front cover of this little book were kindly furnished us by the editor of the above journal.
A. I. ROOT.

www.ingramcontent.com/pod-product-compliance
Lightning Source LLC
Chambersburg PA
CBHW030336170426
43202CB00010B/1145